The History and Development of
NOMOGRAPHY

H.A. EVESHAM

Docent
Press

DOCENT PRESS, LLC
Boston, Massachusetts, USA
www.docentpress.com

Docent Press publishes monographs and translations in the history of mathematics for thoughtful reading by professionals, amateurs and the public.

Cover design by Brenda Riddell, Graphic Details, Portsmouth, New Hampshire.

iv

Preface

There were a number of excellent recitations of the history-to-date of nomography written during the golden age of this mathematical speciality. I may mention, for example, *Contribution a la Théorie et Aux Applications de la Nomographie* by Rodolphe Soreau published in 1902 and *Calcul Graphique et Nomographie* by Maurice d'Ocagne published in 1908. As interest in nomography waned and as the number of mathematicians working in the field became diminishingly small it is understandable that these histories were never updated to capture nomography's golden years.

Thus was the unfinished state of the history of nomography when Harold Ainsley Evesham, at the suggestion of Dr. Eduardo Ortiz of Imperial College, chose the history of nomography as his Doctor of Philosophy thesis topic at the University of London. Evesham's thesis, *The History and Development of Nomography*, completed in 1982 has since become an oft-cited source document on nomography.

The thesis in its original typewritten form circulates in the scholarly community but to the eye unaccustomed to Typit® symbology this rendering can present a hurdle to a full appreciation Evesham's exposition. Thus it seemed that transcribing the work for today's reader would be worthwhile.

Dr. Evesham has overseen this new edition of his thesis. A small number of typographic errors have been corrected and additional details added to a few of the bibliographic references. The hand-drawn figures are those of the original thesis. Where possible higher resolution figures from the historical literature have been used. Only captions appearing in the original manuscript appear in the List of Figures. In some cases the caption in the List of Figures is an abbreviated version of the caption in the text. Consistent with one of the objectives of the republication, mathematical notation has been rendered in TeX [61] and LaTeX [72] following $\mathcal{AMS}$ guidelines [130].

Dr. Evesham has written on the history of nomography in the *IEEE Annals of the History of Computing*, the *Companion Encyclopedia of the History and Philosophy of the Mathematical Sciences*, and *Instruments of Science, An Historical Encyclopedia*. Dr. Evesham was the Assistant Dean of the Department of Science at The University of Luton when he retired.

Finally, I would like to thank Ron Doerfler, the author of *Dead Reckoning: Calculating Without Instruments* for a careful read of the manuscript that contributed to its accuracy.

Scott Guthery, Publisher
Docent Press, LLC
Boston, Massachusetts
www.docentpress.com

Thesis for the Degree of Doctor of Philosophy of the University of London. 1982

Acknowledgements

That a historical study of nomography could be a worthwhile undertaking was first suggested to me by Dr. E.L. Ortiz of Imperial College when I was contemplating a research project on early computational methods. I wish to record my thanks to Dr. Ortiz for this suggestion and for his subsequent advice. This unusual subject has proved to be full of interest. It is unusual in the sense that it lies outside the experience of many present day mathematicians and it is interesting because difficult mathematical problems arise from simple ideas.

The research called for much delving in libraries and a considerable effort in locating exactly where certain documents had been deposited and subsequently retrieving them. For their assistance in this latter task I am most grateful to the Librarian of Luton College of Higher Education and his staff; their help has been invaluable. I also wish to record my appreciation for the help received from the staffs of the British Library, the Science museum Library, the National Maritime Museum, the Royal Artillery Institution, the Institution of Royal Engineers, the School of Military Survey and the Mapping and Charting Establishment, R.E.

When many of the relevant documents were obtained they were, of course, in a foreign language. For most of the translations I must accept responsibility and must be held responsible if subsequently errors are discovered which can be ascribed to faulty translation. In the case of Russian I had to seek outside help. I am grateful to the staff of the translation section of the British Library for their help and I also wish to record my appreciation of the translating skills of Daphne Skillen who translated some of the Russian originals with considerable flair.

Finally, I thank my wife Margaret for her tolerance over the last few years during which period this thesis has been in preparation.

Contents

List of Figures

CHAPTER 1

The Origins of Geometric Computation

Examples of the early use of diagrams as an aid to computation are not difficult to find. Whether they can be considered as early examples of nomography depends upon ones interpretation of the word *nomography*. The interpretation accepted here is that a nomographic method is one which leads to solutions of a class of similar problems rather than to a solution of a single problem. As an illustration of this we may consider the Greek method of solving the equation $x^2 + c^2 = bx$ described by Boyer [11], in which the construction is a specific one for a particular pair of values of b and c. This method cannot be called nomographic. A truly nomographic method would be one which allowed, at the time of reading the diagram, a choice of values for the variables b and c and then gave x. The early precursors of nomograms were invariably concerned with computations related to navigation or astronomy and usually featured a moving element. D'Ocagne cites the Quadratum Horarum Generale of Regiomontanus [1] which appeared in the last quarter of the fifteenth century and which was used to find the solar time at the instant of observation [104].

Many diagrams with moving elements are to be found in Sir R. Dudley's *Del l'Arcano del Mare* which was published in Florence in 1661 and of which a fine copy is to be found at the National Maritime Museum [30]. These diagrams, known as *volvelle* diagrams, are intended for such purposes as "to find the age of the moon" (p.3, f.3) or "to observe and compute the altitude of the Pole Star" (p.23 f.107).

---

[1]Regiomontanus was Johannes Müller who, after the fashion of the times, was known by the Latin form of his birthplace Köinigsberg (King's Mountain), He lived from 1436 − 1476 and amongst other trigonometric achievements seems to have been responsible for the law of sines of spherical triangles. This would confirm his interest in the mathematics of Astronomy and Navigation.

One should consider the examples cited so far as special cases in that they do not arise out of a general theory of geometric computation. An important step in the development of such a theory was made when Fermat and Descartes developed coordinate geometry. However, the step from development to application was a long one, for both developed computational procedures which almost entirely depended upon the straight lines, circles and conics of the older geometry but ignored the multitude of other curves which the new geometry had made available. Furthermore, their methods for determining the roots of equations of the third and fourth degrees suffer from the same failing as does the Greek method for solving the quadratic in that they are specific to an individual equation and not general.

In addition to coordinate geometry certain other ideas needed to become accepted before nomography could begin to establish itself as a useful discipline. The first was that of the graphical representation of data for the purposes of estimation and prediction which depended in part on the development of statistics. Curves for the representation of numerical laws of population and mortality by age appear to have been in use towards the middle of the eighteenth century. The German statistician Pfeiffer is known to have produced such graphical tableaux. Later an application of this type to the statistics of the consumption and maintenance of paving stones in Paris was made by Minard in 1820 in a paper with the title "Plan for canal and railway for the transport of paving stones to Paris" [65]. A second important concept required was that of the representation of three variables on a two dimensional plane. This concept is important since it permits the graphical representation of a double entry table. A double entry table is nothing more than a table of values entered by selecting a row, which designates one of the variables, and a column which designate a second variable. The intersection of this row and column give the value of the third variable. Such a table is the one known as the Table of Pythagoras which gives the result of multiplying two integers. An example is shown in Figure 1.1.

Some special cases of the graphical representation of three variables on a plane are of quite early origin. They arise from geographical or geophysical considerations, two of the variables being the coordinates which fix a position on the surface of the earth and the third representing the measure of some phenomenon at that point. At the beginning of the eighteenth century, Halley recorded lines of equal magnetic declination in this manner and later Euler plotted the line of the magnetic meridian.

1	2	3	4	5	6	7	8	9	10
2	4	6	8	10	12	14	16	18	20
3	6	9	12	15	18	21	24	27	30
4	8	12	16	20	24	28	32	36	40
5	10	15	20	25	30	35	40	45	50
6	12	18	24	30	36	42	48	54	60
7	14	21	28	35	42	49	56	63	70
8	16	24	32	40	48	56	64	72	80
9	18	27	36	45	54	63	72	81	90
10	20	30	40	50	60	70	80	90	100

TABLE 1.1. Table of Pythagoras

The first general application of the idea that three variables could be represented on a two dimensional plane is due to Philippe Buache who, in his "Essai de géographie physique" published in 1752, described how, by taking soundings, one can chart submarine channels and the slope of the sea bed in coastal waters. In particular he did this for the English Channel, recording depths from 10 fathoms increasing by units of 10 fathoms [13]. The idea of level lines representing topographic surfaces is clearly present in this work but Buache does not take it to the logical conclusion of drawing contour lines on land. This step was taken in 1780 by Du Carla of Geneva [29].

The works of Buache and Du Carla were descriptive in nature and in no way represented attempts to calculate anything. For an early application of the idea present in the concept of topographic surfaces to calculation one can note the *Horary Tables* of Margetts published in London in 1790 [79], but the first purely mathematical application is due to Louis Pouchet. The event which was to lead to this application was the decision in France to convert weights and measures to the metric system. Such conversion is greatly helped by the use of double entry tables to convert, for example, the weight of a given quantity of a substance in the old system to the weight of the same quantity in the new system. The authorities did indeed publish such tables but an article of law dated under the Republican calendar as Germinal an IV[2] states "in place of

---

[2]March /April 1795

tables of relationship between old and new measures, which had been provided by the order of the 8th of May, 1790, will be graphical scales to estimate these relationships without having need of any calculation".

Pouchet, who was a member of the Council of Arts and Manufacture, published in 1795 a work on graphical scales for the new weights, measures and monies of the French Republic which included an appendix called "Arithmétique Linéare". In this appendix he gave graphical methods for the elementary arithmetic procedures of addition, subtraction, multiplication and division, for squaring and for the extraction of roots. The method for multiplication is a graphical equivalent of the Table of Pythagoras of Table 1.1 and is given as Figure 1.1.

It will be seen that the curves are the family of equilateral hyperbolæ $xy = c$, where the constant c determines a particular member of the family. Elementary though the idea may be it represents an important conceptual step forward. However, it is not clear whether this advance follows in the footsteps of Buache and Du Carla in the sense that Pouchet has plotted the projection of level lines of a hyperboloid on to a surface or whether the hyperbolæ are merely the results of the variation of the products of two factors. The latter appears to be the case.

An area in which speedy calculations are required is that of ballistics and these calculations would normally have been carried out by means of tables. Graphical procedures, if they are accurate enough, may well produce a faster result with less chance of a mistake. It is therefore no surprise that in the early nineteenth century graphical solutions to problems of ballistics appeared, and it is even less surprising, in view of the development so far, that this should have been in France. In 1814 and 1818 d'Obenheim gave graphical means for solving problems in ballistics ([85] [86]). Level lines were used by Piobert in 1825 in order to verify firing tables for ricochet, which had earlier been calculated by Colonel Lyantey [114]. In 1830, Terquem gave the general principle of graphical double entry tables applied to the graphical construction of Lombard's tables [131]. In the same volume Bellencontre summarizes the works of d'Obenheim on double entry tables as applied to the problems of artillery [6].

Curves of the type $x^2y = $ constant appear in a work by Allix, a naval construction engineer. These curves are used to find, without calculation, weights

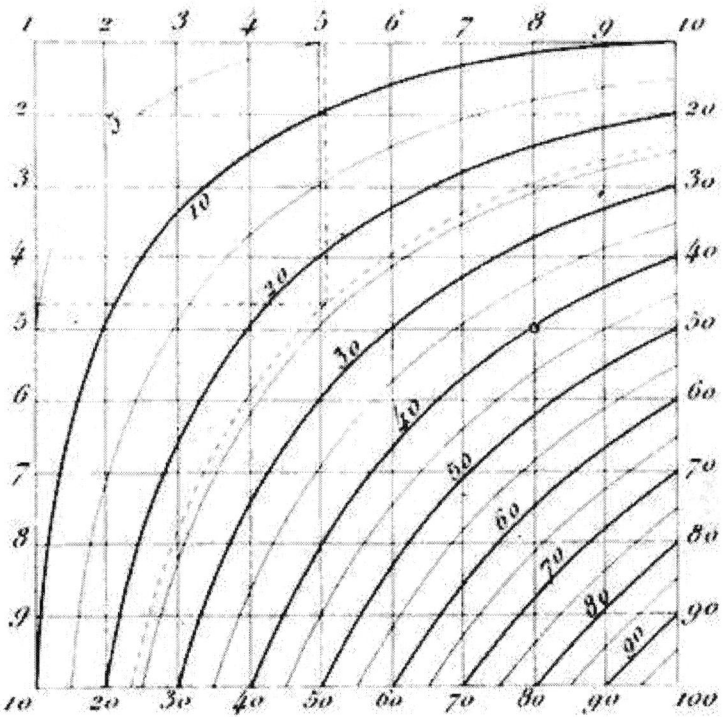

FIGURE 1.1. Pouchet's graphical equivalent of the double entry table of Table 1.1. The hyperbolæ are given by $z_3 = z_1 z_2$. The scale for z_3 is given along the top and down the right hand side.

and measures in the metric system. This work was published in Paris in 1840 [**3**].

With the exception of Piobert's verification of ricochet tables, the examples of the use of level lines given above are of mathematical laws for which an expression in the form of a function of two independent variables was known. The application of the technique to laws resulting from experimental observation are less frequent. One example of this by Capt. Didion appears in the

Journal de l'École Polytechnique. The author represents by curves the results of experiments on the relative accuracy of bullets of different shapes [**27**].

The development of railways in France was a spur to the use of geometric computation. In 1844 Fevre produced a topographic type plan which related the velocity of a locomotive to the weight of the train and the gradient [**44**]. Indeed, railway development had an important effect upon the development of nomography. Lalanne and d'Ocagne in France, and Massau in Belgium, were involved with the state organizations connected with railway construction. The examples used as illustrations in Lalanne's 1846 paper are concerned with railway construction [**65**].

The first case of double entry tables and their corresponding graphical representations appearing together is found in both the French and the English translations of *A Complete Course of Meteorology* by L. F. Kämtz. The French translation appeared in 1843 and contained an appendix by Lalanne in which he gave graphical representations of 42 of the 113 numerical tables contained in the original. The English edition, translated with notes and additions by C.V. Walker, appeared in 1845. In the preface to this work, reference is made to Lalanne as being the first to generalize the representation of three coordinates in one plane. In the appendix, reference is made to what is seen as a consequence of graphical representation, namely, that interpolation can more readily be carried out [**58**].

An important type of nomogram which was developed in the 1880's is the alignment nomogram, the important feature of which is that a line which joins two points on separate scales intersects a third scale at a point which gives the solution to the problem being investigated. This idea had been expressed by Möbius in 1841 for multiplication only, but in two different ways. Firstly, he noted that the line joining the points with ordinates y_1 and y_2 of the parabola $y^2 = x$ cuts the axis of the parabola at a point with abscissa $y_1 y_2$. The second solution he based on the theorem of transversals of Menelaus [**83**].

By the early 1840's, the as yet unnamed subject that was to be called nomography had been conceived. The development of the subject was about to begin in earnest with the work of Lalanne, followed by that of Massau and d'Ocagne.

CHAPTER 2

The Development of a Distinct Discipline

2.1. Lalanne and Anamorphosis

The first important advance to follow the idea of the graphical representation of a double entry table was the consideration of ways by which the construction of such a representation could be improved without affecting its value as a computational tool. This was the principal idea expressed in Lalanne's paper of 1846 [**65**]. Lalanne states that there is no reason why a double entry graphical table should not have the sides of its frame graduated according to some non-regular scale. Referring to a diagram which is essentially the same as Figure 1.1, he points out that if it was to be deformed by a cause such as the unequal contraction of the paper, or if the sketch was molded on to a geometric surface, the accuracy would in no way be altered because the relative position of a point of intersection of a vertical and a horizontal with a particular curve before deformation would not be changed by that deformation; reading such a chart does not depend upon absolute measurement but on relative measurement. He develops this argument by suggesting that there would be advantage in replacing the original curves by curves which were more simple and more easy to construct, in particular by straight lines which could each be fixed by no more than two points.

Recognizing that the deformation described has something analogous to the effects produced by reflection in curved surfaces, to which physicists had already given the name Anamorphosis, Lalanne proposes that the new branch of geometry which he believes must result from these considerations should be given the name of *anamorphic geometry*. In fact the name anamorphosis seems to have been more widely used.

7

The paper referred to here was not the first indication of Lalanne's ideas on anamorphosis. In 1843 a paper had been presented by him to the Paris Academy of Sciences, on the subject of "the substitution of topographic planes for double entry tables" and "a new method of transformation of the coordinates [**64**]" The Commisaires were Cauchy, Élie de Beaumont and Lame, who reported in September 1843 accepting the paper's conclusions [**5**]. In an interesting footnote to their report the commissioners say

> "In effect, supposing that X and Y are functions of x and y respectively, one can generally reduce to the construction of straight lines the solution of an equation of the form
>
> $$f(z) = X\phi(z) + Y\chi(z),$$
>
> $f(z), \phi(z), \chi(z)$ designating three functions of the variable z which one supposes a function of x and of y."

This is intended as an extension to treatment and is taken up by Lalanne in the 1846 paper.

Thus we have clearly expressed by Lalanne in 1843 the idea of anamorphosis, an apparently simple idea but one which would still be the subject of learned papers more than one hundred years later.

Before considering the method by which he proposes to achieve anamorphosis, Lalanne makes some comments on graphical methods. Certainly, some of what he has to say might seem rather trivial today but this is because the construction of graphs is now commonplace. For example, he explains how one may plot a relation of the form $\phi(x, y, z) = 0$ by giving z the successive values of 0, a, $2a$, $3a$, ... and plotting the plane curve relation between x and y in each case. In the development of this theme he acknowledges the influence of Monge's descriptive geometry. In stating the virtues of graphical representation he makes the point that sometimes more information may result than was originally expected and he gives the following example.

The surface area of a cutting or embankment is given by

$$z = \frac{(A \pm y)^2}{2(B \mp x)} - C$$

where y is the axial length and x is the gradient. The depth of the section is given by

$$\frac{A \pm y}{B \mp x}$$

which he notes is $\frac{\partial z}{\partial y}$ and thus, for a constant x, may be interpreted as the slope of a curve. This, however, does seem to be something of a special case.

Another advantage claimed for the graphical method is that it can show properties of a function that are not shown explicitly when the function is written. To illustrate this he takes the example of finding the roots of $x^3 + px + q = 0$, where p and q are both less than one. The approach is to consider x as a parameter and p and q as rectangular Cartesian coordinates. The graph, Figure 2.1, consists of a set of straight lines, each line corresponding to one value of x. Lalanne observes that the envelope of these lines is given by $4p^3 + 27q^2 = 0$ and that the real roots of the equation number one, three with two equal, or three unequal as the point (p, q) lies outside, on, or inside the envelope; i.e., as $4p^3 + 27q^2$ is greater than, equal to, or less than zero. He also notes that the same figure can determine the probability that $x^3 + px + q = 0$ has three real roots when the only knowledge of p and q is that $p < P$ and $q < Q$. This problem is resolved by comparing two areas, one given by the curve $4p^3 + 27q^2 = 0$, the other being $4PQ$.

Lalanne attempts to explain how one may bring about anamorphosis for the given relationship $f(x, y, z) = 0$, but succeeds only in expressing in mathematical symbols what he has already said in words. His argument is as follows:

Given

$$f(x, y, z) = 0 \tag{2.1}$$

and suppose a transformation which expresses (2.1) as

$$F(\phi(x), \psi(y), \pi(z)) = 0 \tag{2.2}$$

then, if we put

$$x_1 = \phi(x), \quad y_1 = \psi(y), \quad \text{and} \quad z_1 = \pi(z) \tag{2.3}$$

(2.2) becomes

$$F(x_1, y_1, z_1) = 0. \tag{2.4}$$

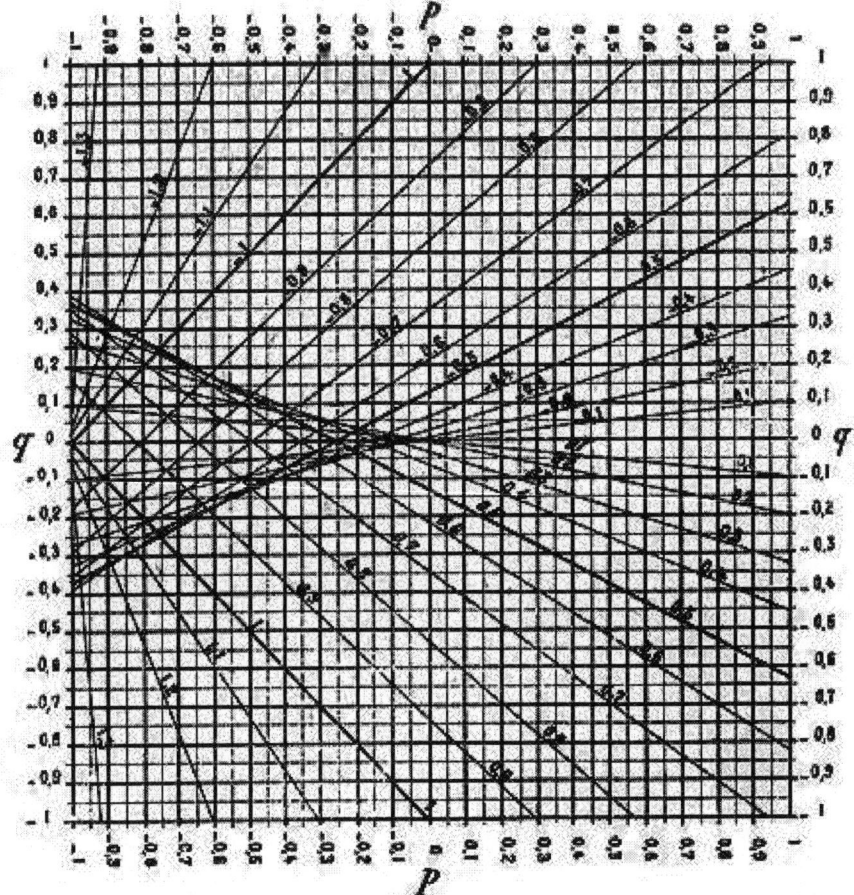

FIGURE 2.1. Lalanne's graphical procedure to find the roots of $x^3 + px + q = 0$. p is measured along the horizontal axis in the range -1 to $+1$, q along the vertical axis in the same range. The numbers on the straight lines are the values of x. It is a curious fact that on Lalanne's original the lines for the x values $\pm 1.1, \pm 1.2, \pm 1.3$ are positioned incorrectly.

If the axes of coordinates are graduated according to $x_1 = \phi(x)$ and $y_1 = \psi(y)$, instead of equal parts then the projections of the z_1 level curves will be straight lines if (2.4) is of the first degree in x_1 and y_1, conic sections if of the second degree and so on. He concludes by pointing out that the degree of (2.4) in x and y can be considerably less than the degree of (2.1) in x and y, a statement with which few would disagree but of little practical use.

The illustrations given by Lalanne of anamorphosis are all rather special in the sense that they owe much to insight and intuition and little to mathematical analysis. The graphical representation of $z = xy$ is transformed from hyperbolæ into straight lines by letting $x_1 = \log x$ and $y_1 = \log y$, with the result that $x_1 + y_1 = \log z$. Lalanne's original diagrams are reproduced in Figure 2.2. The natural extension to functions of the type

$$z = \phi(x)\psi(y) \quad \text{and} \quad z = \phi(x) + \psi(y)$$

is made and the idea of a class of functions with separable variables is expressed. Furthermore, it is pointed out that, even if variables are not separable, they may become so by the substitution $x_1 = \phi(x,y)$, $y_1 = \psi(x,y)$ and the Figure 2.2 is given as illustration. As with many of Lalanne's examples it is taken from his experience as a civil engineer working on railway construction and in this case gives the volume of a cutting.

The expression is

$$z = \frac{ax^2}{x+y}$$

and the suggested substitutions are

$$y_1 = x + y \quad \text{and} \quad x_1 = ax^2$$

giving

$$z = \frac{x_1}{y_1},$$

then, by fixing z successively as α, 2α, 3α, ..., we get

$$y_1 = \frac{1}{\alpha}x_1, \quad y_1 = \frac{1}{2\alpha}x_1, \quad y_1 = \frac{1}{3\alpha}x_1, \ldots;$$

i.e., a set of straight lines passing through the origin of the coordinates x_1, y_1.

Without giving any details Lalanne advocates the use of projective transformations in conjunction with anamorphosis to deduce "an infinity of other

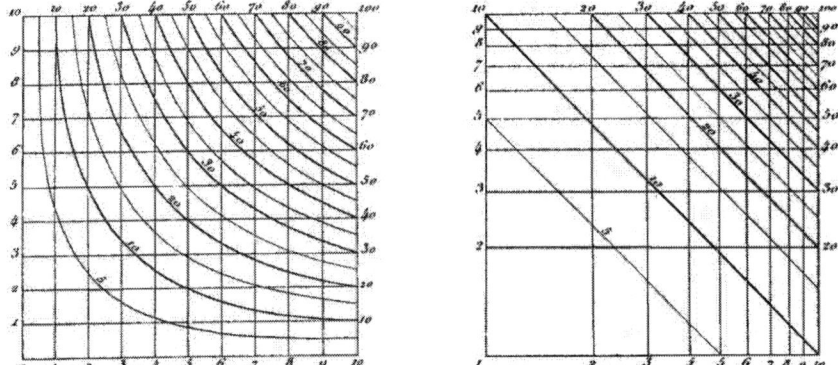

FIGURE 2.2. Lalanne's illustration of anamorphosis. The graph on the left is of $z = xy$, the values of z being written on the hyperbolæ, while that on the right is of $\log z = \log x + \log y$.

analogous figures." By doing so he is anticipating something which has become an important feature of nomography.

Lalanne points out a few of the mathematical consequences of anamorphosis. The most obvious of these is that the substitution $x_1 = \log x$ will lead to a shift of the origin since $x_1 = 0$ will correspond to $x = 1$. He also points out that there is no reason to suppose that the functions $\phi(x)$ and $\psi(y)$, used to graduate the axes, should increase or decrease in a constant manner and that it is possible that they have a maximum or a minimum. To illustrate he uses $y = a + bx + cx^2$ and applies anamorphosis to reduce it to a straight line by the substitution $x_1 = a + bx + cx^2$. The straight line is thus $y = x_1$, but the parabola is not represented by the whole of this line for, if we assume for the purpose of illustration that $b > 0$ and $c > 0$ then $x_1 = a + bx + cx^2$ will have a minimum when $x = \frac{-b}{2c}$; i.e., when $x_1 = a - \frac{b^2}{4c}$.

A further point that he makes, and which is worthy of note, is that for a function z of x and y in which the two independent variables are separable, anamorphosis is not unique. Returning to the example $z = xy$ and the anamorphosis already considered which produced $z_1 = x_1 + y_1$ he points out that a further anamorphosis applied to the straight lines can produce concentric circles. The anamorphosis required is to $x_2 = \sqrt{x_1}$, $y_2 = \sqrt{y_1}$, and $z_2 = \sqrt{z_1}$, giving

$$z_2 = x_2^2 + y_2^2;$$

fixing z_2 in turn as $\alpha, \beta, \gamma, \ldots$ we have the concentric circles

$$x_2^2 + y_2^2 = \alpha^2, \quad x_2^2 + y_2^2 = \beta^2, \quad x_2^2 + y_2^2 = \gamma^2, \ldots$$

Taking up the general point made by Cauchy in 1843 [5], Lalanne contents himself by taking the expression

$$f(z) = X(x)\phi(z) + Y(y)\psi(z),$$

and after making the substitutions

$$x_1 = X(x), \quad y_1 = Y(y)$$

and pointing out that, for fixed values of z,

$$f(z) = x_1\phi(z) + y_1\psi(z)$$

are straight lines.

Lalanne also makes the important point that anamorphosis is not confined to Cartesian coordinates but can be applied to polar coordinates or indeed to any coordinate system and he illustrates the point with reference to the hyperbolic spiral $\rho\omega = a^2$ which, with the substitutions $\rho_1 = \log\rho$ and $\omega_1 = -\log\omega$, becomes an Archimedean spiral.

So convinced is Lalanne of the merits of geometric computation that he advocates a Universal Calculator to replace the slide rule, the use of which is so common in England. His Universal Calculator is reproduced as Figure 2.3. Amongst the calculations which can be carried out using it are multiplication and division; raising to the powers of 2, 3, and higher powers and finding the corresponding roots; multiplication and division by 2π; the calculation of πr^2, $(4/3)\pi r^3$; simplification of calculations containing g, $(1/2)g$, $2g$, and $\sqrt{2g}$; and the solution of ratios to find chemical equivalents. The advantages that he sees for his device over the slide rule include the following. Results depend only on reading, there being no moving parts, and any shrinkage or deformation cannot influence the result (presumably shrinkage, or deformation of a slide rule may be of part of the rule only); there are physical difficulties over the use of the rule which he claims is not really portable while the chart is portable and cheap. In these views, and others that he expresses, he does seem to have abandoned rationality in order to advance his idea. His final comment on the Universal Calculator is that he looks forward to the time when it will appear in school rooms and in public squares alongside clocks and sundials assisting in calculations as the clock and sundial assist in the measurement of time.

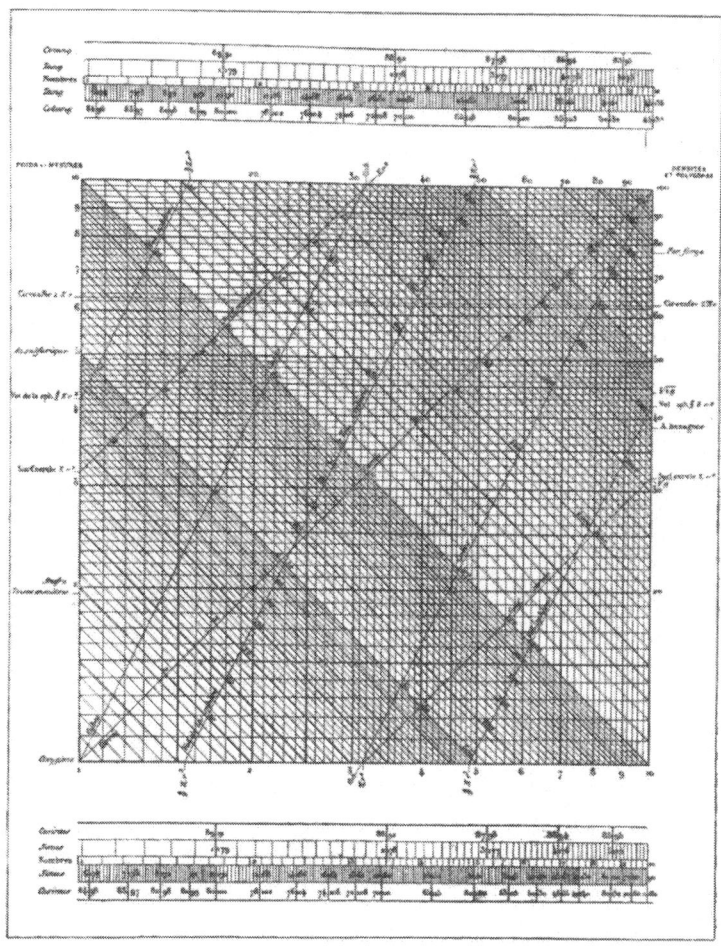

FIGURE 2.3. Lalanne's Universal Calculator as it appeared in his 1846 paper [65]

Lalanne includes in his paper a short section on the application of graph-ical representation to certain natural laws, by which he means laws governing population size and mortality. The point that he makes in this connection is that ignorance of the explicit form of the function of a natural law need not inhibit the construction of a graph. To illustrate this point I take a simplified version of his example.

Suppose, for some species of living creature in a well defined area, we have the following table giving the number living at each age.

Age	0	1	2	3	4	5
No. living	0	100	50	40	30	20

TABLE 2.1.

The problem is to find the number of individuals between the ages of a and a_1 where $a_1 > a$.

If x is the number of individuals between 0 and a and y is the number of individuals between 0 and a_1 then, if z is the number of individuals between a and a_1, $z = y - x$, which will be represented by straight lines although $x = f(a)$ and $y = f(a_1)$ are unknown. x and y can be tabulated according to Table 2.2.

Age	0	1	2	3	4	5
x (or y)	0	10	100	190	220	240

TABLE 2.2.

Lalanne's chart, Figure 2.4, is constructed in the following way. Two un-scaled perpendicular lines of equal length are drawn, one for the x axis the other for the y axis. The z lines are then constructed; $z = 0$ is the line joining the free ends of the x and y axes and the other z lines are parallel to it and at suitable intervals to provide a regular scale for z. The y scale is now marked using the information given by the table. The x scale is identical to the y scale.

To find the number of individuals between two ages one takes the horizontal through the lower age, the vertical through the higher age and takes the value of the sloping line on which they intersect. Interpolation is possible.

By 1846 many profound ideas on geometric computation had been expressed by Lalanne. The ideas had not been explored in great depth but nevertheless they constituted a body of knowledge which, on the one hand provided useful computational tools, and on the other gave a foundation for nomography on which those who followed could build.

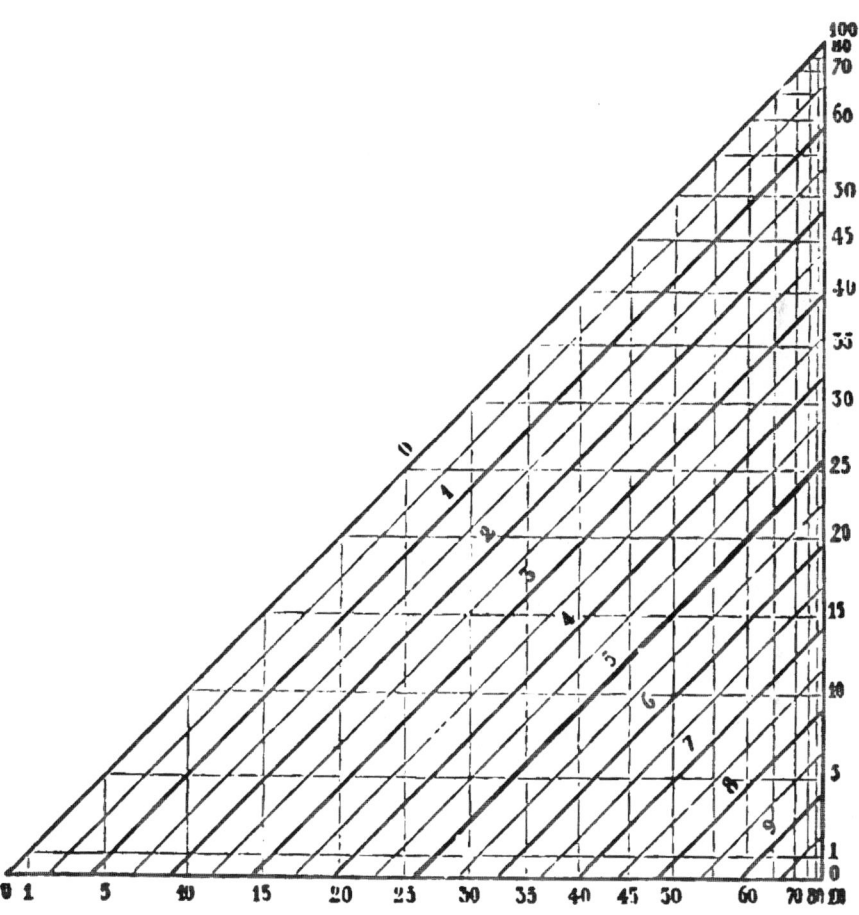

FIGURE 2.4. Lalanne type chart for a natural law in a case for which the explicit form of the law is not known.

2.2. Saint-Robert's Criterion

At the meeting of the Academy of Sciences of Turin on April 7, 1867, Paul de Saint-Robert read a paper which represented the first truly mathematical attempt to determine whether a given equation $F(x, y, z) = 0$ could be transformed into an equation of the form $Z(z) = X(x) + Y(y)$, thus facilitating anamorphosis [122]. Ironically, in view of Lalanne's slightly contemptuous attitude to the slide rule, Saint-Robert was led to his analysis by consideration of a slide rule.

Saint-Robert was engaged in editing tables for the calculation of difference in altitude based on the variations of barometric pressure and temperature when he conceived the notion of a slide rule, analogous to the logarithmic rule, by means of which he could obtain mechanically the required results. He wrote

> "In reflecting on this reduction of double entry tables into a slide rule, I saw that one can, in certain cases, solve certain equations with three variables by means of a slide rule graduated in a convenient manner."

He states the problem in the following way. Given the equation in three variables

$$F(x, y, z) = 0 \qquad (2.5)$$

is it possible to construct three parallel scales, two fixed and one moveable, in such a manner that in any position of the three scales the corresponding values satisfy equation (2.5)? If the three scales AB, CD and EF are positioned as in Figure 2.5, EF being the moveable scale and a, b, c, d specific points on the scales, then $Cd = Aa + bc$ or, letting $X = Aa, Y = bc$ and $Z = Cd$, then

$$Z = X + Y. \qquad (2.6)$$

If we suppose that Z is a function of z only, X of x only and Y of y only then, in order that the slide rule may be used to solve the given equation, it is necessary that the two equations (2.5) and (2.6) give the same value of z for the same pair of values of the independent variables x and y.

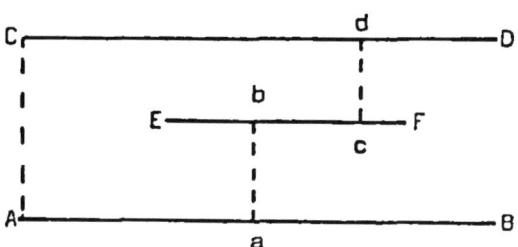

FIGURE 2.5. Three Parallel Axes

Thus, the problem he sets out to solve is this: Given an equation in three variables

$$F(x, y, z) = 0 \qquad (2.7)$$

can it be transformed into an equation

$$Z(z) = X(x) + Y(y) \qquad (2.8)$$

such that the values of z given by both equations are equal for the same pair of values of x and y?

In his solution, Saint-Robert firstly observes that which Lalanne had noted, that there is no problem in the case of $z = xy$, which reduces to

$$\log z = \log x + \log y$$

nor in the case of

$$\phi(z) = \psi(x)\chi(y)$$

which reduces to $Z = X + Y$ when

$$Z = \log \phi(z), \quad X = \log \psi(x), \quad \text{and} \quad Y = \log \chi(y).$$

The barometric pressure formula with which he was concerned is of the type shown above, for it can be written as

$$\frac{Z}{A}\left(\frac{2 - az}{1 - az}\right) = (274 + x)\left(\frac{y_0}{y} - 1\right)$$

and transformed into $Z = X + Y$ by

$$Z = \log\left(\frac{Z}{A}\frac{2 - az}{1 - az}\right), \quad X = \log(274 + x), \quad \text{and} \quad Y = \log\left(\frac{y_0}{y} - 1\right).$$

He notes in passing that he has constructed a device on these lines, the *rhab-dohypsologiste*.

For those forms of equation (2.7) for which the reduction to the form (2.8) is not apparent by inspection, or after simple rearrangement, Saint-Robert finds a condition which, if satisfied, shows that reduction is possible and gives a method of achieving this reduction.

Starting with equation (2.8) and noting that x and y are independent variables but that z is a function of both x and y, he proceeds as follows. Given

$$Z(z) = X(x) + Y(y) \tag{2.9}$$

and differentiating it partially with respect to x,

$$\frac{\partial Z}{\partial z}\frac{\partial z}{\partial x} = X'$$

and with respect to y,

$$\frac{\partial Z}{\partial z}\frac{\partial z}{\partial y} = Y'$$

we have

$$\frac{\partial Z}{\partial z} = \frac{X'}{\frac{\partial z}{\partial x}} = \frac{Y'}{\frac{\partial z}{\partial y}}.$$

If $R = \frac{x'}{y'}$, then

$$R = \frac{\partial z}{\partial x}\bigg/\frac{\partial z}{\partial y}. \tag{2.10}$$

An expression for R can also be obtained from (2.7); i.e., $F(x, y, z) = 0$. Differentiating partially with respect to x gives

$$\frac{\partial F}{\partial x} + \frac{\partial F}{\partial z}\frac{\partial z}{\partial x} = 0;$$

i.e.,

$$\frac{\partial z}{\partial x} = -\frac{\partial F}{\partial x}\bigg/\frac{\partial F}{\partial z}$$

and with respect to y,

$$\frac{\partial F}{\partial y} + \frac{\partial F}{\partial z}\frac{\partial z}{\partial y} = 0;$$

i.e.,

$$\frac{\partial z}{\partial y} = -\frac{\partial F}{\partial y} \Big/ \frac{\partial F}{\partial z} .$$

Therefore,

$$R = \frac{\partial F}{\partial x} \Big/ \frac{\partial F}{\partial y} .$$

Returning to the equation $R = \frac{x'}{y'}$ and taking logarithms to the base e of both sides we have

$$\ln R = \ln X' - \ln Y';$$

differentiating partially with respect to x gives

$$\frac{\partial \ln R}{\partial x} = \frac{1}{X'} X''$$

and again with respect to y gives

$$\frac{\partial^2 \ln R}{\partial x \partial y} = 0.$$

Thus, Saint-Robert's criterion is as follows:

An equation $F(x, y, z) = 0$ can be reduced to the form

$$Z(z) = X(x) + Y(y)$$

if R satisfies the condition

$$\frac{\partial^2 \ln R}{\partial x \partial y} = 0$$

where

$$R = \frac{\partial F}{\partial x} \Big/ \frac{\partial F}{\partial y} \quad \text{or} \quad R = \frac{\partial z}{\partial x} \Big/ \frac{\partial z}{\partial y} .$$

If this condition is satisfied, $Z(z), X(x)$ and $Y(y)$ may be found as follows:

- For $X(x)$, integrate $\frac{\partial \ln R}{\partial x} = \frac{X''}{X'}$ twice.
- For $Y(y)$, integrate $Y' = \frac{X'}{R}$ which contains no x.
- For $Z(z)$, substitute X and Y in (2.8) and use (2.7) to eliminate x and y.

Saint-Robert concludes his paper with two well chosen examples, well chosen because a knowledge of higher mathematics, in the one case of hyperbolic functions and in the other of elliptic functions, enables the reader to check the correctness of the method.

Saint-Robert's criterion is important because it is the first step in a line of enquiry which many others followed, and because it gives an effective procedure for solving the problem, if it can be solved, which leaves nothing to insight or intuition.

As an illustration of this, consider the case of $z = xy$ referred to so often already. Here

$$F(x, y, z) = z - xy = 0$$

so that

$$\frac{\partial F}{\partial x} = -y, \qquad \frac{\partial F}{\partial y} = -x, \qquad R = \frac{y}{x}$$

$$\ln R = \ln y - \ln x, \qquad \frac{\partial \ln R}{\partial x} = -\frac{1}{x},$$

and

$$\frac{\partial^2 \ln R}{\partial x \partial y} = 0,$$

so Saint-Robert's criterion is satisfied. Therefore,

$$\frac{X''}{X'} = -\frac{1}{x}, \qquad \ln X' = -\ln x, \qquad X' = \frac{1}{x}, \qquad X = \ln x$$

and

$$Y' = \frac{1}{x}\frac{x}{y} = \frac{1}{y}, \qquad Y = \ln y$$

so that

$$Z = \ln x + \ln y = \ln xy = \ln z$$

and finally

$$\ln z = \ln x + \ln y.$$

The constants of integration have not been ignored; they cancel out.

2.3. The Contribution of Junius Massau

The year 1884 has an important place in the history of nomography. It was the year during which Maurice d'Ocagne published his first paper on the subject, a paper which introduced alignment nomograms and therefore marked the beginning of a new phase. It was also the year during which a Belgian engineer, Junius Massau, published the results of his work, but in his case the work was concerned with intersection nomograms and followed naturally from the work of Léon Lalanne. Therefore, in one year two papers appeared, one of which considerably enhanced existing knowledge while the other branched along an entirely novel path. 1884 is also an important year in a more general way. Before 1884 publications on, or related to, graphical computation were occasional and demonstrated good ideas rather than contributions to a growing body of knowledge, but after 1884, and in particular from 1884 to 1932, there was a steady flow of papers on both practical and theoretical aspects and this period must be regarded as the most important for nomography. In passing, it is of interest to note that the subject seems to have been dormant between 1932 and 1956 in the sense that there were apparently no steps taken to develop it, but that from 1956 onwards there has been a renewed interest in it. This renewed interest takes two main forms, one the application of computing and approximation techniques and the other the application of mathematics to the problems of anamorphosis and the superposition of functions. Throughout the period from 1884 to the present, however, nomograms have been in continuous use whether or not the development of the subject was dormant.

Massau's paper is one which is full of interesting ideas, some of which have survived still bearing his name, others have an anonymous presence in the literature or are attributed to others, while some have disappeared completely. Massau was a civil engineer who was concerned with the construction of railways in Belgium just as Lalanne had been in France. His paper was published by the association of former students of the Écoles Speciales de Gand (Ghent) and was part of a series of articles under the general heading "L'intégration graphique et ses applications" which appeared at intervals between 1878 and 1900 [82]. It is the sections numbered 177 to 207 which are of particular relevance to this study. The ideas on which the work is based have origins earlier than 1884, for a sub-heading of the paper reads "Développement des théses présentés au concours universitaire de 1873-74."

A part of Massau's work which is well known, is on the expression of the equation $F(x, y, z) = 0$ in the form

$$Z_1(z)X(x) + Z_2(z)Y(y) = 1.$$

I defer consideration of this to a later section where I also examine a similar exercise by Lecornu which appeared two years after Massau's.

Massau begins the relevant section of his paper with a review of the work of Lalanne, which is clearly the starting point for his own work. He points out that Lalanne's methods are inconvenient in that they require much time for the construction of charts. He quotes an engineer named Ricour who, in order to produce four charts for a particular problem concerned with railway construction, required 56 hours of calculation and 112 hours of drawing; i.e., one whole week's work. Massau's work, therefore, has a practical object, to make improvements which will reduce the total time required for the construction of charts. To this end he must look to Lalanne's concept of anamorphosis.

Perhaps the most important contribution of Massau to the development of nomography was the introduction of determinants into the discipline, although at that time he may only have been using the tools of a more skilled mathematician. He presented a theoretical argument which would be the cause of much mathematical activity in the future, raising a problem which will be considered at length later. Under the heading "Use of a general system of co-ordinates," he poses the problem of representing the variable w defined by the equation

$$f(u, v, w) = 0 \tag{2.11}$$

Without explaining how this could be done, he requires the construction of two sets of curves

$$f_1(x, y, u) = 0 \tag{2.12}$$

$$f_2(x, y, v) = 0 \tag{2.13}$$

In a rectangular Cartesian system having co-ordinates (x, y) equation (2.12) will represent curves, each one of which is attached to a particular value of u; similarly (2.13) will represent v curves. Thus to each point of the (x, y) plane there will correspond a pair of values, one u, one v, which can be considered as curvilinear co-ordinates at that point. Therefore equation (2.11) can be used to obtain a set of curves representing w, expressed in the curvilinear co-ordinates u and v, by fixing particular values for w. To obtain the equation for w in the rectangular Cartesian co-ordinates x and y it is only necessary to eliminate u and v between equations (2.11), (2.12) and (2.13).

Massau now supposes that u, v, and w can be represented by straight lines, in which case,

- for the u lines $ax + by + c = 0$, where a, b, and c are functions of u;
- for the v lines $a'x + b'y + c' = 0$, where a', b', and c' are functions of v;
- for the w lines $a''x + b''y + c'' = 0$, where a'', b'', and c'' are functions of w.

On eliminating x and y he obtains the condition,

$$\begin{vmatrix} a & b & c \\ a' & b' & c' \\ a'' & b'' & c'' \end{vmatrix} = 0.$$

This can be written as

$$\begin{vmatrix} a(u)/c(u) & b(u)/c(u) & 1 \\ a'(v)/c'(v) & b'(v)/c'(v) & 1 \\ a''(w)/c''(w) & b''(w)/c''(w) & 1 \end{vmatrix} = 0,$$

a form occasionally referred to as the *Massau determinant*.

Massau observes that this determinant contains six functions and is more general than the form considered by Lalanne which only contained four functions. He has in mind here the form suggested to Lalanne by Cauchy which can be written

$$Z_1(z)X(x) + Z_2(z)Y(y) = 1.$$

He further points out that his method contains the latter, for if the u lines are given by $x = \phi(u)$ and the v lines by $y = \psi(u)$, then

$$\begin{vmatrix} \phi(u) & 0 & 1 \\ 0 & \psi(v) & 1 \\ A(w) & B(w) & 1 \end{vmatrix} = 0$$

giving

$$\frac{A(w)}{\phi(u)} + \frac{B(w)}{\psi(v)} = 1$$

which is of the Cauchy form.

Massau continues his paper with rather brief notes on topics which are so commonplace in later nomography that one accepts them as commonsense, giving little thought to their origins. He notes Lalanne's idea of graphical elimination, that is, if one is given

$$f(u, v, t) = 0,$$
$$f_1(u, t, t') = 0,$$
$$f_2(t, t', w) = 0$$

then to find the relationship between u, v and w it is not necessary to eliminate analytically t and t' but merely to plot the three relationships on the same sheet and carry out the elimination by the suppression of the curves t and t'. He also suggests the use of transparent sheets, each carrying one set of curves, being superimposed on a sheet carrying another set of curves, there being sufficient common variables to make this possible.

Massau devotes considerable space to the discussion of systems of straight lines. He defines the degree of a system of lines as follows:

> If the lines have for their equation $ax + by + c = 0$, where a, b and c are functions of u, then, if the three functions of u are each of integer degree n, the system is of degree n.

The discussion with which he follows this concerns itself with systems of the first and second degrees and, although it is not amongst the most durable of Massau's contributions, it is worth some attention for its interesting approach.

In a u-system of the first degree represented by $ax + by + c = 0$ it may be supposed, for convenience that $a = Au - A_1$, $b = Bu - B_1$, $c = Cu - C_1$, where A, A_1, B, B_1, C, and C_1 are constants. Then $ax + by + c = 0$ can be written

$$u(Ax + By + C) = A_1x + B_1y + C_1$$

or

$$u\alpha = \beta.$$

It will be observed that if $u = 0$ then $\beta = 0$ and if $u = \infty$ then $\alpha = 0$. Massau then suggests taking a new system of coordinates based on the axes CX and CY in which CX is the line $\beta = 0$, and so corresponds to the value $u = 0$, and CY is the line $\alpha = 0$ and corresponds to the value $u = \infty$. Since the equation

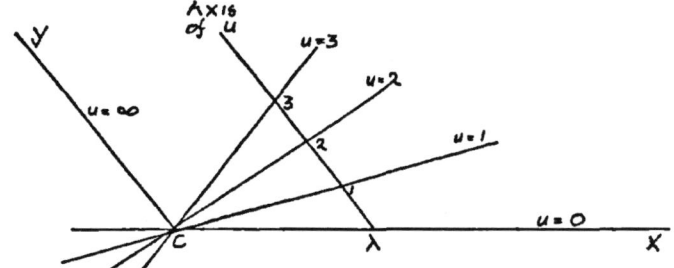

FIGURE 2.6. The intersection of a straight line with the lines of a straight line system

for every u line in terms of α and β is $u\alpha = \beta$, it is clear that every u line passes through the new origin C. In general, the equation of the new system is $\lambda Y = uX$ and the lines can easily be determined by making $X = \lambda$ and $Y = u$. in fact the straight line $X = \lambda$ carries the scale of u which Massau calls the axis of u (Figure 2.6).

A particular concern of Massau is the intersection of any straight line with the lines of a straight line system. Suppose that the system is

$$\lambda Y = \mu X \tag{2.14}$$

and the straight line is given by

$$X = X_0 + K\mu' \tag{2.15}$$
$$Y = Y_0 + M\mu' \tag{2.16}$$

where μ' is the distance along this line from the point (X_0, Y_0).

On substituting (2.15) and (2.16) into (2.14), we have

$$\lambda(Y_0 + M\mu') = \mu(X_0 + K\mu');$$

i.e.,

$$K\mu\mu' + X_0\mu - \lambda M\mu' - \lambda Y_0 = 0$$

or more generally, $Auu' + Bu + Cu' + D = 0$ which is of the form

$$u' = \frac{Mu + n}{M'u + N'}.$$

This last form leads to the theorem given by Massau that any straight line is cut by a u-system of the first degree in a homographic scale of u.

The problem of the representation of a function by three systems of the first degree is given some attention. If the variables are u, v and w then, by a suitable change of axes, one set, say the w-set, can have the form

$$y - \frac{wx}{\lambda} = 0.$$

Let the u lines have the form:

$$ay + bx = c$$

and the v lines the form

$$a'y + b'x = c'$$

where a, b and c are first degree functions of u and a', b' and c' first degree functions of v.

If we eliminate x and y from these three we have

$$\frac{w}{\lambda} = \frac{b'c - bc'}{ac' - a'c} \tag{2.17}$$

which has the form:

$$w = \frac{A + Bu + Cv + Duv}{A' + B'u + C'v + D'uv}. \tag{2.18}$$

Massau now poses a more difficult and more practical problem; given (2.18) how can we draw a chart for w by means of three systems of straight lines of the first degree?

Since a comparison of (2.17) with (2.18) is not possible, as (2.17) contains more coefficients than (2.18), Massau resorts to an interesting alternative.

In (2.17), w will be indeterminate in the case where $c = c' = 0$. For w to be similarly indeterminate in (2.18) we must have

$$A + Bu + Cv + Duv = 0$$

and

$$A' + B'u + C'v + D'uv = 0.$$

These two equations can be combined to form a second degree equation in either u or v. Suppose that the roots are real and that u_0 and v_0 are a pair of solutions. With the substitutions

$$u - u_0 = u', \quad v - v_0 = v', \quad c = u', \quad \text{and} \quad c' = v',$$

we then have

$$\frac{w}{\lambda} = \frac{b'u' - bv'}{av' - a'u'} \tag{2.19}$$

and

$$w = \frac{B_1 u' + C_1 v' + D u' v'}{B_1' u' + C_1' v' + D' u' v'} \tag{2.20}$$

which can be written

$$w = \frac{(B_1' + Ev')u' + v'(C_1 + (D - E)u')}{(B_1' + E'v')u' + v'(C_1' + (D' - E')u')}. \tag{2.21}$$

Equations (2.19) and (2.21) can now be compared and if we choose

$$E = 0, \quad E' = 0, \quad \text{and} \quad \lambda = 1,$$

then the three systems of lines are

$$y - wx = 0,$$
$$C_1' y - C_1 x + u'(D'y - Dx - 1) = 0,$$
$$\text{and} \quad B_1' y - B_1 x + v' = 0$$

in which C_1, C_1', B_1, and B_1' can be found from the given form, and u' and v' from the roots of the quadratic equation in u or v.

Massau also attacks the preceding problem by means of trilinear co-ordinates. Trilinear co-ordinates are a form of homogeneous co-ordinates related to a fixed triangle ABC, the triangle of reference (Figure 2.7). The co-ordinates (α, β, γ) of a point P are such that α is the perpendicular distance from P from the side BC, β that from AC, and γ that from AB. α, β and γ have the form $p - x\cos\theta - y\sin\theta$. The sides AB, BC, and CA therefore correspond to $\gamma = 0$, $\alpha = 0$, and $\beta = 0$ respectively. It is clear that only two of the co-ordinates are required to determine a point, the third can be found from the relationship $a\alpha + b\beta + c\gamma = 2\Delta$, where $a = BC, b = AC, c = AB$ and Δ is the area of triangle ABC.

Massau does not say why he has chosen to use trilinear co-ordinates but it is a most appropriate choice for first degree systems since, as we have already

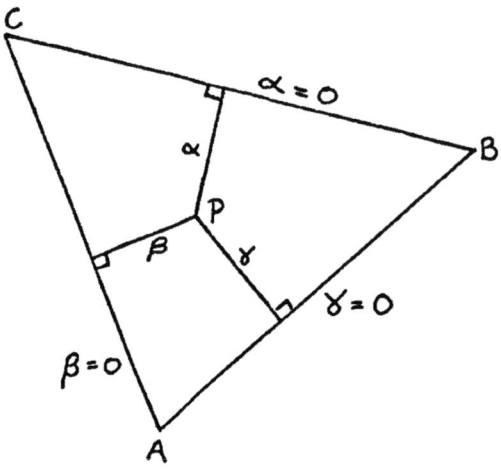

FIGURE 2.7. Trilinear System

noted, all lines of a first degree system intersect in one point. Three such systems will, therefore, in general, give rise to three points which can be used as vertices for the triangle of reference, as in Figure 2.8.

An equation of the form $b\beta - c\gamma = 0$, in which b and c are of the first degree in u, represents a straight line for any specified value of u. Furthermore, it is always satisfied by $\beta = 0$ and $\gamma = 0$ showing that all of these lines pass through A. Similar reasoning applies to the v lines which converge on B and to the w lines which converge on C. We have then the following equations:

$$u \text{ lines:} \qquad b\beta = c\gamma,$$
$$v \text{ lines:} \qquad c'\gamma = a'\alpha,$$
$$w \text{ lines:} \qquad a''\alpha = b''\beta.$$

If we eliminate from these three equations the co-ordinates $\alpha, \beta,$ and γ we have

$$a''c'b = a'b''c.$$

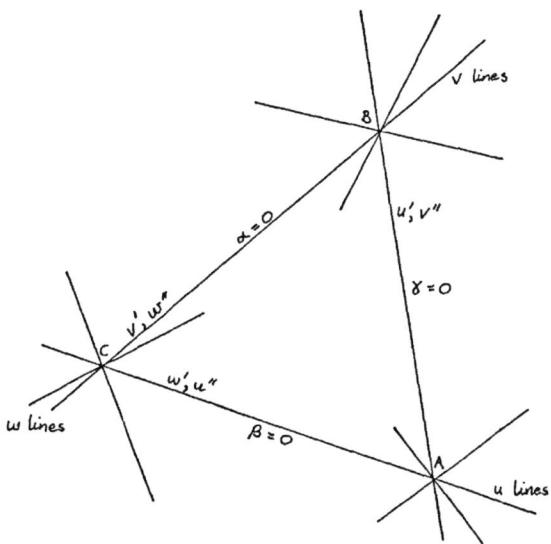

FIGURE 2.8. Three Point System

It should be noted that each side of the triangle of reference will correspond to a line from two of the systems; AB is both a u line, say u', and a v line, say v''. Similarly, we can say that BC corresponds to v' and w'' and AC to u'' and w'.

It follows that if $u = u'$ and $v = v''$, then w is indeterminate since an infinity of w lines intersect the straight line of which AB forms a part.

To make use of these results, Massau applies them in the case of

$$D + Au + Bv + Cw + A'vw + B'wu + C'uv + Euvw = 0 \qquad (2.22)$$

which can be seen to be a form equivalent to (2.18).

In order that w should be indeterminate it is necessary that

$$D + Au + Bv + C'uv = 0 \qquad (2.23)$$

and

$$C + A'v + B'u + Euv = 0. \qquad (2.24)$$

We can eliminate v between (2.23) and (2.24), getting

$$u^2(AE - B'C') + u(AA' + DE - BB' - CC') + A'D - BC = 0. \qquad (2.25)$$

The problem is now resolved quite simply.

Equation (2.25) is solved, assuming that it can be, for real u. The two roots u' and u'' are assigned, one to the side AB of the triangle of reference, say u', and one to the side AC, u''. Equation (2.23) will now give corresponding values for v; viz.,

v'', corresponding to u', is attributed to the side AB,
v', corresponding to u'', is attributed to the side BC.

The values w' and w'' are obtained by expressing that which makes v indeterminate. One such equation for this is

$$D + Au + Cw + B'uw = 0. \qquad (2.26)$$

Substituting u' in (2.26) will give w'' which is assigned to the side BC and then u'' gives w' assigned to the side CA.

The triangle of reference then gives two lines of each system. A system needs three lines to be completely determined. To achieve this it is necessary to take a set of values satisfying (2.22). For example, set

$$u = 0, \quad v = 0, \quad \text{and} \quad w = -\frac{D}{C},$$

and choose any point M to represent these.

Then the u-system is represented by AM, AB, and AC corresponding to $u = 0$, $u = u'$, and $u = u''$ (Figure 2.9). The v- and w-systems are similarly represented by three straight lines. It should be noted that the triangle of reference and the point M can be freely chosen so that the resulting chart is the most convenient.

This particular approach is of some interest for in 1907 d'Ocagne published a paper describing his concept of critical points [**99**]. Although he was concerned with alignment nomograms the same idea of indeterminacy based on a triangle is used. It is unfortunate that d'Ocagne, who seems always to have been concerned with claiming priority for his own ideas, did not give Massau the credit which

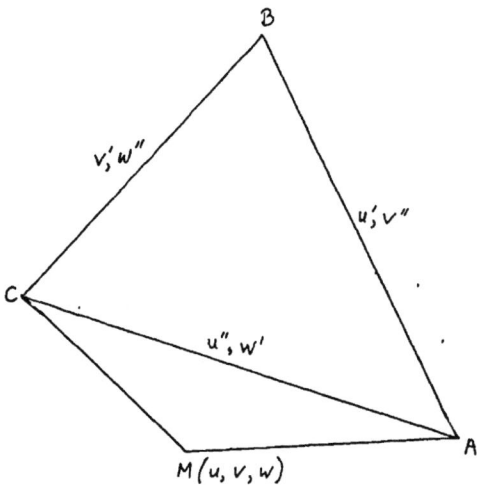

FIGURE 2.9. Three Line System

was his due. Of course, it may have been that d'Ocagne had not read Massau's paper himself and was familiar only with its major results, but this seems unlikely.

Massau treats second degree systems also by means of trilinear co-ordinates. The equation of a u line is

$$l\alpha u^2 + m\beta u + n\gamma = 0 \tag{2.27}$$

in which l, m,and n are constants.

The line $u = 0$ corresponds to $\gamma = 0$; i.e., AB in the triangle of reference and, if we write (2.27) in the form

$$l\alpha + m\frac{\beta_1}{u} + n\frac{\gamma_1}{u^2} = 0,$$

we see that when $u^{-1} = 0$; i.e., when u is infinite, $\alpha = 0$ so that the line $u = \infty$ corresponds to BC of the triangle of reference.

If we differentiate (2.27) with respect to u we have

$$2l\alpha u + m\beta = 0 \tag{2.28}$$

and if we substitute

$$l\alpha u = -\frac{m\beta}{2},$$

obtained from (2.28), in (2.27), we have

$$\frac{-m\beta}{2}u + m\beta u + n\gamma = 0;$$

i.e.,

$$m\beta u + 2n\gamma = 0. \tag{2.29}$$

Equations (2.28) and (2.29) each represent straight lines which intersect in a point on the envelope of (2.27). The equation to the envelope is obtained by eliminating u between (2.28) and (2.29) and is

$$m^2\beta^2 - 4ln\alpha\gamma = 0$$

which is the condition for (2.27) to have equal roots. The equations (2.28) and (2.29) give the points of contact of the lines $u = 0$ and $u = \infty$ with the envelope, for, when $u = 0$ and $\gamma = 0$, we have $\beta = 0$ or the point A and, when $u = \infty$ and $\alpha = 0$, we have $\beta = 0$ or the point C.

Massau states the theorem that each straight line $u = u_0$ of a system is cut by the others in a homographic scale of u of which the points are found on an auxiliary system of the first degree given by $lu\alpha + m\beta$ where $U = u + u_0$. Furthermore, the u line parallel to u is cut in a scale proportional to u and can be used as the axis of u.

Lines of a second degree system will not all intersect in one common point as do those of a first degree system. Thus, such a system may be represented by the whole lines shown in Figure 2.10.

The points of intersection referred to in the theorem are marked with circles. They must satisfy the two equations,

$$lu^2\alpha + mu\beta + n\gamma = 0 \quad \text{and} \quad lu_0^2\alpha + mu_0\beta + n\gamma = 0.$$

If we subtract these we have

$$l(u + u_0)\alpha + m\beta = 0 \tag{2.30}$$

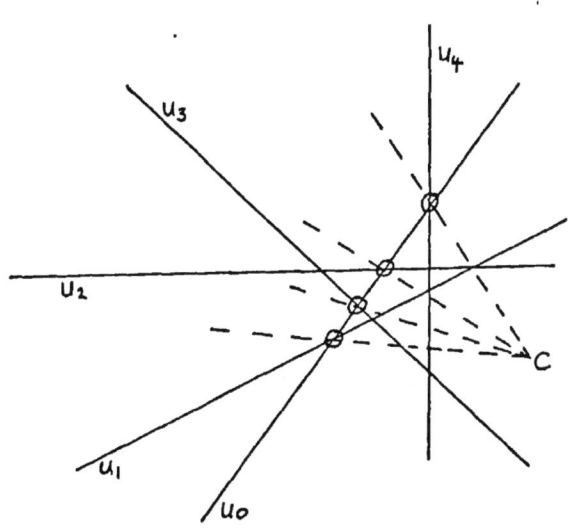

FIGURE 2.10. Lines of a second degree equation

or writing $U = u + u_0$ we have

$$lU\alpha + m\beta = 0.$$

This is the equation of a first degree system in U, a system which converges on the vertex C of the triangle of reference. It has already been shown that such a system would cut a straight line in a homographic scale of U, and, since u_0 is a constant in each case, in a homographic scale of u. This proves the first part of Massau's theorem. The second part follows from earlier work in which the axis of u is shown to be a line parallel to $u = \infty$ and the line BC is known to correspond to $u = \infty$.

The system can be determined analytically in quite an easy manner in the case where three particular tangents to the envelope are given. If the tangents are CB corresponding to $u = \infty$, with point of contact C, AB corresponding to $u = 0$ with point of contact A and DE corresponding to $u = u_0$ then the trilinear co-ordinates α, β, and γ are known since the triangle of reference is known.

It is necessary to find l, m, n for the relationship

$$lu^2\alpha + mu\beta + n\gamma = 0.$$

We suppose that the equation of DE is given in the form

$$P\alpha + Q\beta + R\gamma = 0.$$

This must be identical with

$$lu^2\alpha + mu_0\beta + n\gamma = 0$$

and therefore the equation to any line may be written

$$P\frac{u^2}{u_0}\alpha + Q\frac{u}{u_0}\beta + R\gamma = 0$$

which determines analytically the system.

It should be noted from earlier results that being given the tangent $u = \infty$ amounts to being given an axis for u, being given the line $u = 0$ amounts to being given the origin for u and being given the line $u = u_0$ is sufficient to find the scale of u (Figure 2.11).

It follows then that if on a tangent to a conic a scale for u is set out starting from any origin, and if, from the points on this scale tangents to the conic are drawn, a system of the second degree is obtained. If, on another tangent, a scale of v is set out, then another system of the second degree is obtained. Furthermore, a tangent will be, at the same time, a line of u and a line of v. There will be some relation between u and v and it is easy to see what it is since the v line will be cut by the u lines according to a homographic scale of u. A relationship of the form

$$Auv + Bu + Cv + D = 0$$

must apply, since it can be written in the form

$$v = -\frac{Bu + D}{Au + C}.$$

From the foregoing it can be seen that the construction of a second degree system need not be difficult. For example, given the equation in the form $lu^2\alpha + mu\beta + n\gamma = 0$, the envelope is given by $m^2\beta^2 - 4ln\alpha\gamma = 0$. Given

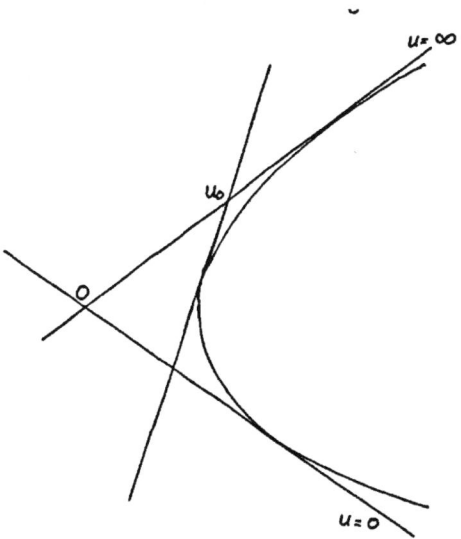

FIGURE 2.11

two tangents and their points of contact one can make these points of contact C and A of the triangle of reference and assign to the tangents the values of $u = \infty$ and $u = 0$ respectively. They intersect at the point B of the triangle of reference (Figure 2.12).

The line (2.30); i.e., $l(u + u_0)\alpha + m\beta = 0$ becomes $2lu\alpha + m\beta = 0$ when $u_0 = u$, a first degree system converging on C which will give the points of intersection of the second degree system with $u = 0$; i.e., AB. If we make $\alpha = 0$ in $lu^2\alpha + mu\beta + n\gamma = 0$ we have $mu\beta + n\gamma = 0$, a first order system converging on A, which will give the points of intersection of the second degree system with $u = \infty$; i.e., BC. We then have two points for each u line. Lines other than $u = \infty$ and $u = 0$ could have been selected.

Finally, we note that if one of α, β, or γ is constant we have particular forms of the envelope. If α is constant the envelope

$$m^2\beta^2 = 4ln\gamma\alpha$$

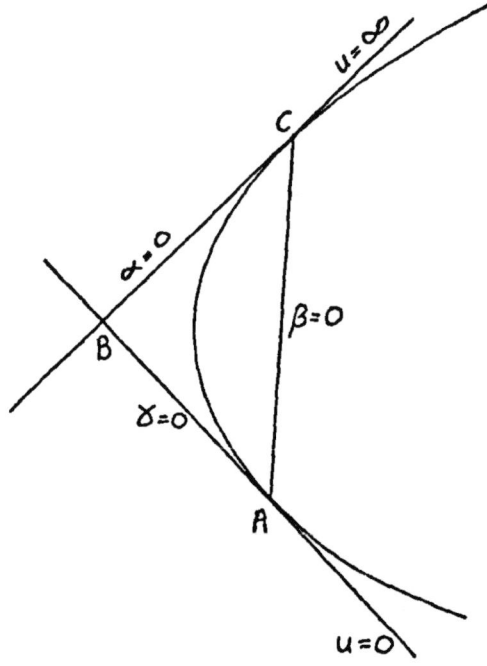

FIGURE 2.12

becomes a parabola, if β is constant it becomes a hyperbola with $\alpha = 0$ and $\gamma = 0$ as asymptotes and if γ is constant it becomes another parabola.

I have not encountered the use of trilinear co-ordinates in connection with nomography outside Massau's paper and it seems likely that the idea was never fully developed. Indeed, the evidence suggests that his paper may have been more frequently referred to than read. The only indication that anyone has taken this part of it seriously is the fact, already mentioned, that d'Ocagne appears to have taken Massau's ideas on indeterminate values and adapted them to his own work.

After the digression into trilinear co-ordinates, Massau's paper returns to more conventional ideas. He considers the case of a function which can be represented by three systems of the second degree. He returns to the point which he made early in the paper that three systems of straight lines depend upon a relation of the form

$$\begin{vmatrix} a & b & c \\ a' & b' & c' \\ a'' & b'' & c'' \end{vmatrix} = 0$$

but in the case of second degree systems a, a', and a'' represent functions of the second degree in u, v, and w. It follows that the determinant will give an equation of the sixth degree containing u, v, w, u^2, v^2, and w^2 to the first degree.

The real problem is the inverse problem and Massau is the first writer to express it. He states it in the particular form to which his work has led him, that is, given an equation of the sixth degree in u, v, w, u^2, v^2, and w^2, how can one construct a computation chart? He recognizes that the general method would consist of identifying the given equation with

$$\begin{vmatrix} a(u) & b(u) & c(u) \\ a'(v) & b'(v) & c'(v) \\ a''(w) & b''(w) & c''(w) \end{vmatrix} = 0$$

Massau states that this leads to laborious calculations which is both a perceptive recognition of the nature of the problem and an understatement of its difficulties. He does, however, treat some particular forms.

First he considers the case

$$w = \frac{a + bu + cu + duv}{a_1 + b_1 u + c_1 v + d_1 uv}. \tag{2.31}$$

The method is to take rectangular Cartesian axes Dx and Dy and to put

$$x = bu + cv + duv \tag{2.32}$$

$$y = b_1 u + c_1 v + d_1 uv. \tag{2.33}$$

Equation (2.31) now becomes

$$w(y + a_1) = x + a$$

which shows that the w lines form a system of the first degree.

We can eliminate from (2.32) and (2.33) v and u in turn, giving

$$\frac{x - bu}{c + du} = \frac{y - b_1 u}{c_1 + d_1 u} \tag{2.34}$$

$$\frac{x - cv}{b + dv} = \frac{y - c_1 v}{b_1 + d_1 v}. \tag{2.35}$$

Equation (2.34) represents a second degree system in u and equation (2.35) second degree system in v. The three systems can easily be constructed. However, there is an anomaly with which Massau deals at length. It arises from the fact that if from (2.34) and (2.35) one calculates x and y one does not return to (2.32) and (2.33); a denominator is present which leads to an extraneous solution of form

$$Auv + Bu + Cv + D = 0.$$

Values of u and v satisfying this solution render w indeterminate. Massau indicates how to deal with this problem.

His second example is

$$w^2 + w\frac{N}{M} + \frac{P}{M} = 0 \tag{2.36}$$

in which M, N, and P are functions of the first degree in u, v, and uv. The method proposed is to put

$$x = \frac{N}{M} \tag{2.37}$$

$$\text{and} \quad y = \frac{P}{M}. \tag{2.38}$$

These transform (2.36) into $w^2 + wx + y = 0$, a system of second degree straight lines. Again, from (2.37) and (2.38), we can eliminate u, then v, to get the v and u lines, both of which are systems of the second degree. In this case also an extraneous solution presents itself. Massau treats this second example a second time and in more depth by making use of determinants because, as he states, the calculations become simple by so doing. The importance of this lies not in the particular example but in the use of determinants.

The brief description given here of Massau's contribution to the development of nomography by no means covers the whole of the material in his paper for there is much of a practical nature concerned with civil engineering. However, the purpose of this thesis is to trace the development of the ideas present in nomography and the study must be limited to that.

Massau's contribution to nomography is considerable. His paper bristles with ideas, some of which have not stood the test of time, but those that have occupy respected positions within the discipline.

2.4. The First Papers of Maurice d'Ocagne

When d'Ocagne published his first paper on nomography he was twenty-two years old and a student engineer; the citation refers to him as Élève-Ingénieur des Ponts et Chaussées [88]. The paper describes "A new method of graphical calculation" and attacks a problem previously investigated by Lalanne, (Figure 2.1). D'Ocagne approaches the problem in an entirely different manner.

The problem is that of finding the solution of an equation of the type

$$x^n + px + q = 0.$$

D'Ocagne concentrates in his later development on the special cases of $n = 2$ and $n = 3$ which are of the most practical value. However, the method is developed for a general n. His approach is graceful, one is almost tempted to say beautiful, in its simplicity. It could today be of value to anyone wishing to solve a large number of cubic equations, since it provides good approximations to the roots which can then be improved using a computing device. I have not come across any case of it being so used.

The basis of d'Ocagne's method is the use of an unusual coordinate system called by him *parallel coordinates*. It will be necessary to digress to describe briefly this system. For a full account of the system one can do no bettor than read a set of articles by d'Ocagne also published in 1884, in which he describes two simple systems of tangential coordinates, the parallel coordinates already referred to, and axial coordinates [87]. Parallel coordinates owe their origin to the line coordinates of Plücker [115].

The basis of the system of parallel coordinates is a pair of parallel lines AU, BV and a transversal AB. AB need not be perpendicular to the parallel lines but in this simple description it will be assumed to be.

In Figure 2.13 the coordinates u and v are measured from A along AU and B along BV respectively, positive values upwards and negative values downwards. Thus the pair of coordinates (u, v) representing the points M and N, where $AM = u$ and $BN = v$, define completely the straight line MN.

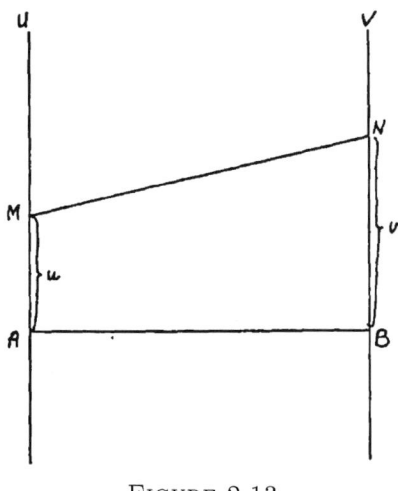

FIGURE 2.13

Certain consequences follow from such a system, those relevant to the discussion are:

i. If the coordinates of a variable straight line (u, v) are connected by a relation $F(u, v) = 0$, then the variable straight line is a tangent to a certain curve, the envelope, having $F(u, v) = 0$ as its equation.

ii. If $F(u, v) = 0$ is of the first degree then the envelope reduces to a point; i.e., the equation $v + au + b = 0$ represents a point P.

iii. If through P a line is drawn parallel to the axes AU and BV to cut AB at Q, then $\frac{QA}{QB} = \frac{1}{a}$ (Figure 2.14).

iv. From (iii) it follows that P is between AU and BV if a is positive and outside if a is negative.

v. If AP cuts BV in B' then $BB' = -b$. Similarly, if BP cuts AU in A', then $AA' = \frac{-b}{a}$ (Figure 2.15).

The above results are applied to the solution of the equation $X^n + pX + q = 0$ as follows. Choose as variables p and q and represent p by u and q by v, also replace X by a particular value α.

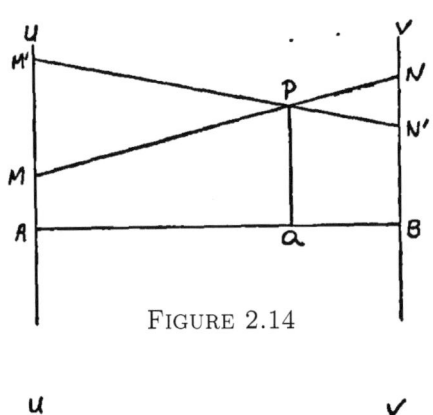

FIGURE 2.14

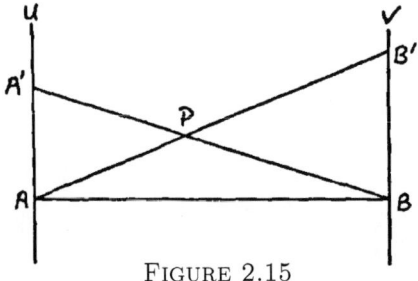

FIGURE 2.15

The equation is now

$$v + \alpha u + \alpha^{\eta} = 0$$

which represents a point P (in contrast to Lalanne's treatment in which the corresponding equation represents a line). This point is easy to construct by virtue of (v) above. We only need to find the point of intersection of AB' and BA' where $AA' = -\alpha^{n-1}$ and $BB' = -\alpha^{n}$ As a check, or alternative to one of the lines, we have the result of (iii) that

$$\frac{QA}{QB} = -\frac{1}{\alpha}.$$

Taking a sequence of values of α all positive in order that the points lie between the parallel axes, points corresponding to the values α_1, α_2, α_3, ... of the parameter are obtained. As each point is obtained the corresponding value of α should be recorded near to that point.

These points will lie on a certain curve C_n which can be drawn when sufficient points have been plotted. Figure 2.16 shows the curve for C_3 which accompanied d'Ocagne's original paper; the same general form applies for other values of n. Note in particular that (iii) shows that when $\alpha = 1$ the point is always mid-way between the parallel axes.

In order to find the roots of an equation consider the example of Figure 2.16, $x^3 - 7x + 6 = 0$. We see, by comparing it with $v + \alpha u + \alpha^n = 0$ that $n = 3$, α corresponds to x, $u = -7$ and $v = +6$, so we must align the point -7 on the left hand parallel axis (p on the figure) with $+6$ on the right hand parallel axis (q on the figure). The line joining these points intersects C_3 at 1 and 2 and these are two roots of the equation. There must now be a third root which, in view of the sign of 6, must be negative, therefore in the equation $x^3 - 7x + 6 = 0$, replace x by $-x$ giving $x^3 - 7x - 6 = 0$. Solving this in the manner indicated above we find that there is one root, namely $+3$. Therefore the third root of $x^3 - 7x + 6 = 0$ is -3.

This method is the first example of an *alignment nomogram* although neither word is used in the paper to describe it. The superiority of the method of estimating values by aligning two points over the method of estimating the point of intersection of three lines is undeniable. Great credit is due to d'Ocagne for introducing the method.

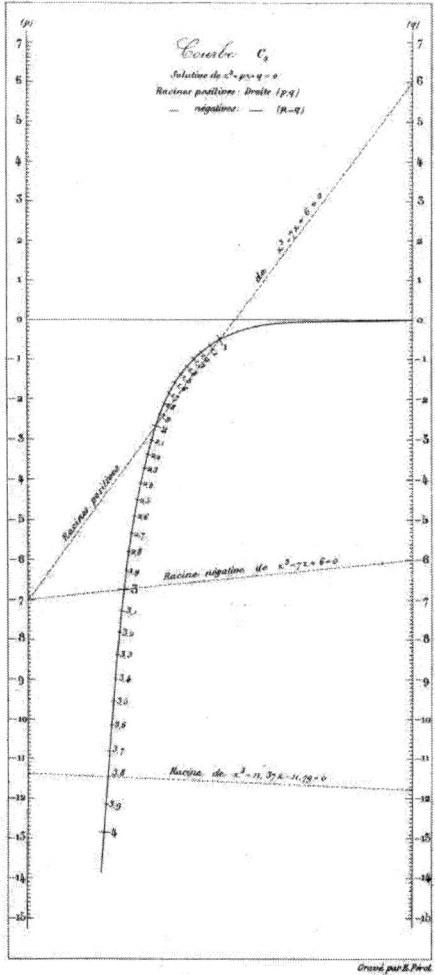

FIGURE 2.16. D'Ocagne's alignment nomogram for $x^3 + px + q = 9$. Its simplicity is appreciated by a comparison with Lalanne's nomogram of Figure 2.1.

2.5. Lallemand's Hexagonal Nomogram

In 1886 a paper was presented to the Paris Academy of Sciences by Charles Lallemand in which he described his hexagonal nomogram [69]. This type of nomogram originated in 1883 when Lallemand, employed by the Nivellement Général de la France, was preoccupied with the simplification of the calculations carried out by that organization. He was later to become the director of the Nivellement Général. A brochure describing the method appeared in 1885 but this was for use within the organization only, as d'Ocagne makes clear in his *Nomographie* [90]. In a paper on the origins and state of nomography presented to the Academy of Sciences by Lallemand in 1922, the method is referred to with some pride [71].

The basis of the hexagonal method is the following property of geometry – I translate from Lallemand's 1806 paper.

> "The algebraic sum of the projections of a segment of a line on two axes having an angle of 120° between them is equal to the projection of the same segment onto the internal bisector of the angle between these axes."

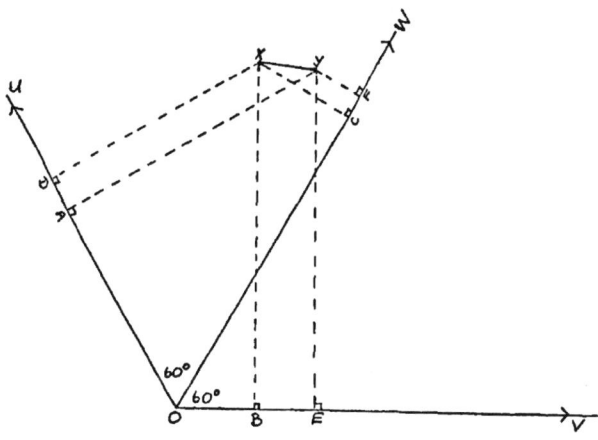

FIGURE 2.17

In Figure 2.17, OU and OV are the axes with OW as their internal, bisector. If XY is the given line then the result states that $AD + BC = DC$. A simpler form of the result is obtained by making OX the straight line, in which case Figure 2.18 applies and the result is $OA + OB = OC$. The proof of this is elementary.

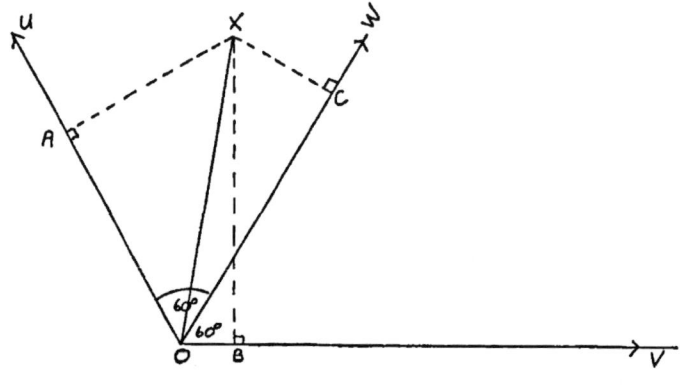

FIGURE 2.18

For the use of Figure 2.18 as a nomogram OU, OV, and OW can be scaled according to the laws

$$f_1(u), \quad f_2(v), \quad \text{and} \quad f_3(w)$$

respectively, in the directions indicated, to give the relationship

$$f_1(u) + f_2(v) - f_3(w) = 0.$$

To read the nomogram, given say u and v, it is necessary first to find X, the point of intersection of the perpendiculars to OU and OV at u and v respectively. Then from X drop a perpendicular onto OW to find the corresponding value of w. Lallemand suggests the use of an oriented transparency which takes the form of a regular hexagon on which is engraved the three diameters, which he refers to as index lines. If the point of intersection of these diameters is placed over X, then XA, XB, and XC can be made to coincide with parts of the diameters and the task of reading the nomogram will be greatly eased. Of course, it will also be necessary to engrave some parallel sets of lines, perpendicular to the index lines, in order to orientate the transparency.

The method is of interest partly because it appears to be independent of earlier work on nomograms and partly because it is the first example that requires an orientated transparency. Other writers had already suggested a transparent sheet carrying an engraved line as an aid to reading a nomogram, but the reason for this was to keep the nomogram clean rather than as an essential part of it.

There is more flexibility in the method than may be obvious from a cursory inspection. The lines OU, OV, and OW may be displaced in directions perpendicular to themselves without changing the positions on the lines of A, B, and C in Figure 2.18. Figure 2.19 shows that the three lines may be displaced parallel to themselves which implies that 0 need not be the origin of the three variable axes; in this case A, B, and C represent the corresponding origins for u, v, and w. Such flexibility means, for example, that if it is convenient to do so, the three scales can be the sides of an equilateral triangle, or that if the range of the variables is to be increased, then it can be done without unduly increasing the size of the nomogram by displacing the scales to accommodate the increased range. In Figure 2.19 the increase in the range is shown by the broken lines.

Lallemand claimed more for his nomogram. He claimed that it was applicable to all equations in which, directly or after anamorphosis, the variables could be separated into groups, of at the most two, in a sum of products of functions such that

$$\sum f_1(x_1, y_1) f_2(x_2, y_2) f_3(x_3, y_3) \ldots = 0.$$

The method by which he proposes to deal with such an expression is simply to replace the corresponding linear scales by diagrams having two sets of isopléthes. The word *isopléthés* occurs frequently in nomographic writings of this time; it designates a curve having a fixed value for some parameter. The more variables the more complicated is the nomogram and in this respect Lallemand's idea is more limited than he himself believed. As an illustration of what was done consider Lallemand's own hexagonal nomogram giving the deviation of a compass for a particular ship, Le Triomphe.

The formula with which it deals is

$$\delta = a + m \sin \zeta + n \cos \zeta + b \sin 2\zeta + c \cos 2\zeta,$$

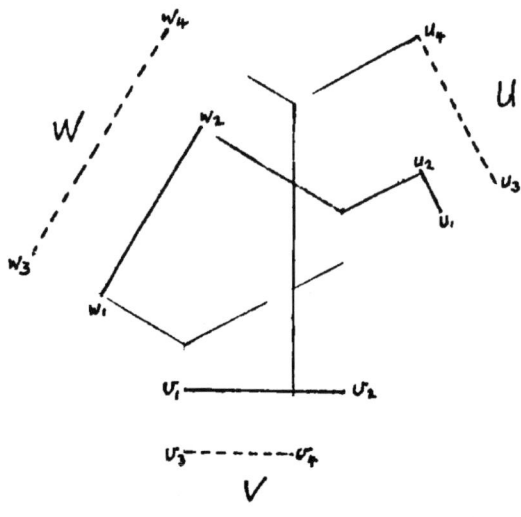

FIGURE 2.19

where ζ is the compass bearing; $a, b,$ and c are constants particular to the ship; $m,$ and n are known functions of Θ (the magnetic declination), and H (the horizontal magnetic component).

It is split up as follows:
$$w = \delta,$$
$$v = m \sin \zeta,$$
$$\text{and} \quad u = n \cos \zeta + b \sin 2\zeta + c \cos 2\zeta + a,$$

giving $w = v + u.$

Θ and H do not appear explicitly on the nomogram as they are obtained from the latitude (λ) and longitude (L) and therefore a network for (λ, L) is substituted for a network for (Θ, H).

Figure 2.20 shows the plan of the nomogram, the basis of which is the equilateral triangle ABC, and Figure 2.21 is Lallemand's final nomogram. The

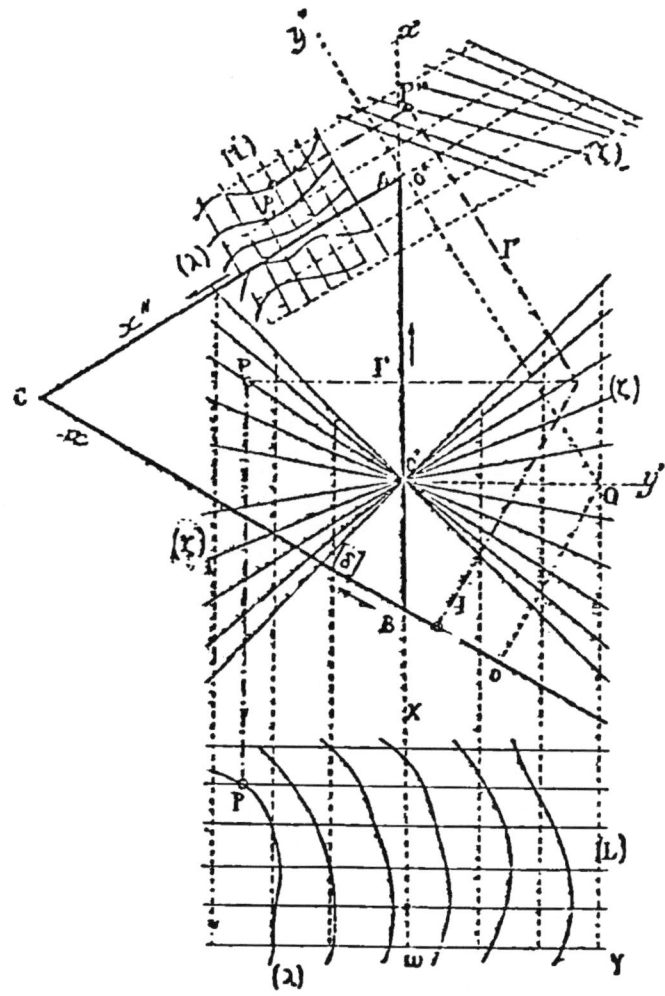

FIGURE 2.20. Plan for Lallemand's nomogram of Figure 2.21

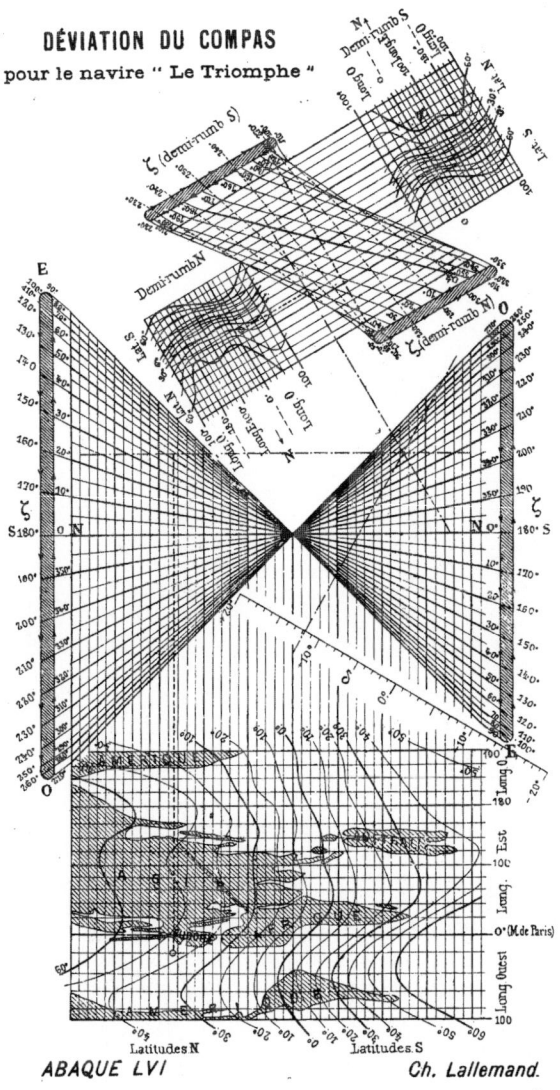

FIGURE 2.21. An example of Lallemand's hexagonal nomogram

dotted lines indicate an example where $\lambda = 42°N$, $L = 20°W$, and $\zeta = 41.5°$ giving $\delta = 11.8°$.

Lallemand has certainly taken the simple idea to a very high level of completion.

2.6. Two Early Attempts to Solve a Problem of Anamorphosis

It has already been noted that Saint-Robert investigated the problem of expressing the equation $F(x, y, z) = 0$ in the form $Z(z) = X(x) + Y(y)$ and produced a criterion to determine when this is possible with a method for finding $Z(z), X(x)$, and $Y(y)$ when the criterion is satisfied. Although Saint-Robert does not appear to have been concerned with the application of his result to nomography, it has stood the test of time. As illustration of this one can cite two relatively recent examples in which it is considered worthy of mention; in *Nomography* by Edward Otto published in 1963 ([**107**], [**108**]), and in an article by Džems-Levi published in 1959 [**40**].

Two further attempts at a similar problem were made in 1884 and 1886. Both were concerned with putting the equation $F(x, y, z) = 0$ into the form

$$Z_1(z)X(x) + Z_2(z)Y(y) = 1 \qquad (2.39)$$

where z is considered to be a function of both x and y. One may wonder why the form (2.39) was selected rather than the form chosen by Saint-Robert. The form (2.39) is rather more general in that it is a simplification of

$$Z_1(z)X(x) + Z_2(z)Y(y) + Z_3(z) = 0.$$

Also, Cauchy and his colleagues when considering Lalanne's 1843 paper had remarked that the form $f(z) = X\phi(z) + Y\chi(z)$ could "generally be reduced to the construction of straight lines."

The first attack on the problem, in 1884, was by the Belgian engineer Janius Massau [**82**]. In looking at his work we denote partial derivatives in the usual manner,

$$p = \frac{\partial z}{\partial x}, \quad q = \frac{\partial z}{\partial y}, \quad r = \frac{\partial^2 z}{\partial x^2}, \quad s = \frac{\partial^2 z}{\partial x \partial y}, \quad \text{and} \quad t = \frac{\partial^2 z}{\partial y^2}.$$

Firstly, Massau obtains a value R from these partial derivatives which in turn will have been calculated from the original equation $F(x, y, z) = 0$. His R is given by,

$$R = \frac{rq}{p} - 2s + \frac{tp}{q}.$$

Two quantities U_1, a function of x, and U_2, a function of y, are introduced and linked to the required quantities X and Y by

$$\frac{X''}{X'} = U_1, \quad \text{and} \quad \frac{Y''}{Y'} = U_2.$$

By partial differentiation he is able to obtain from (2.39) the equation

$$pU_2 + qU_1 = R. \tag{2.40}$$

Three more equations are obtained from (2.40) by suitable partial differentiation. They are

$$rU_2 + sU_1 + qU_1' = \frac{\partial R}{\partial x}, \tag{2.41}$$

$$sU_2 + pU_2' + tU_1 = \frac{\partial R}{\partial y}, \tag{2.42}$$

$$\frac{\partial r}{\partial y}U_2 + rU_2' + \frac{\partial s}{\partial y}U_1 + \frac{\partial q}{\partial y}U_1' = \frac{\partial^2 R}{\partial x \partial y}. \tag{2.43}$$

Equations (2.40), (2.41), (2.42), and (2.43) will give the values of U_1, U_2, U_1', and U_2'.

From the way in which U_1 and U_2 are defined, the following conditions must apply:

$$\frac{\partial U_1}{\partial y} = 0 \quad \text{and} \quad \frac{\partial U_2}{\partial x} = 0. \tag{2.44}$$

Also from (2.39), by suitable partial differentiation, the relationship

$$\frac{Z_1}{Z_2} = \frac{p}{q}\frac{Y'}{X'} \tag{2.45}$$

is obtained.

The method is applied as follows. Form the following equations from the given $F(x, y, z) = 0$:

$$p\lambda + q\mu = R,$$

$$r\lambda + s\mu + q\nu = \frac{\partial R}{\partial x},$$

$$s\lambda + p\pi + t\mu = \frac{\partial R}{\partial y}, \tag{2.46}$$

$$\text{and} \quad \lambda\frac{\partial r}{\partial y} + \mu\frac{\partial s}{\partial y} + r\pi + \nu\frac{\partial q}{\partial y} = \frac{\partial^2 R}{\partial x \partial y}.$$

Use the equations to find λ and μ. The conditions

$$\frac{\partial \lambda}{\partial x} = 0 \quad \text{and} \quad \frac{\partial \mu}{\partial y} = 0 \tag{2.47}$$

must be satisfied. Then we have

$$\frac{X''}{X'} = \mu \quad \text{and} \quad \frac{Y''}{Y'} = \lambda$$

which will give X and Y.

The relationship between Z_1 and Z_2 is now obtained from (2.45); i.e.,

$$\frac{Z_1}{Z_2} = \frac{p}{q}\frac{Y'}{X'}$$

and the functions Z_1 and Z_2 from the form (2.39).

It is possible that the coefficient matrix of (2.46) will have a rank less than four, as is the case in the example given below in which the rank is three. In this case one variable may, in theory, be chosen arbitrarily but, in practice, some care will be required if the conditions $\frac{\partial \lambda}{\partial x} = 0$ and $\frac{\partial \mu}{\partial y} = 0$ are to be satisfied.

Consider the case
$$F(x, y, z) = z - x^2 y^2 = 0.$$
Then
$$p = 2xy^2, \quad q = 2x^2 y, \quad r = 2y^2, \quad t = 2x^2, \quad s = 4xy, \quad \text{and} \quad R = -4xy.$$
We have the equations
$$y\lambda + x\mu = -2,$$
$$y\lambda + 2x\mu + x^2\nu = -2,$$
$$2y\lambda + x\mu + y^2\pi = -2,$$
$$\text{and} \quad 4y\lambda + 4x\mu + 2x^2\nu + 2y^2\pi = -4.$$
This system reduces to
$$y\lambda + x\mu = -2,$$
$$x\mu + x^2\nu = 0,$$
$$\text{and} \quad y\lambda + y^2\pi = 0.$$
In order that the first equation may be satisfied it is possible to choose λ as a function of y only and μ as a function of x only by taking $\lambda = y^{-1}$ and $\mu = x^{-1}$. These give $\nu = -x^{-2}$ and $\pi = -y^{-2}$ but we do not require these.

The conditions
$$\frac{\partial \lambda}{\partial x} = 0 \quad \text{and} \quad \frac{\partial \mu}{\partial y} = 0$$
are satisfied. We have
$$\frac{X''}{X'} = \frac{-1}{x} \quad \text{and} \quad \frac{Y''}{Y'} = \frac{-1}{y}$$
leading to $X = \ln x$ and $Y = \ln y$. Then
$$\frac{q}{p} = \frac{Z_2 Y'}{Z_1 X'}$$
leads to $Z_2 = Z_1 = Z$, and $XZ_1 + YZ_2 = 1$ gives
$$Z = \frac{1}{\ln xy} = \frac{2}{\ln z}.$$
Therefore the form is
$$2\ln x + 2\ln y = \ln z.$$

The second attempt was in 1886 and is due to Léon Lecornu, a French mining engineer [**76**]. Lecornu begins his paper with a reference to Lalanne's work on graphical tables and anamorphosis and poses the problem of finding when a given relation between three variables $F(x, y, z) = 0$ can be put into the form

$$a_z X(x) + b_z Y(y) + c_z = 0,$$

where a_z, b_z, and c_z are functions of z and z is a function of x and y.

We have an additional clue to his reason for starting with this form. He quotes from the French edition of Culmann's *Traité de Statique Graphique* [**22**] thus, "On ne peut donner de régle générale pour transformer $F(x, y, z) = 0$ en $a_z x' + b_z y' + c_z = 0$" (here x' is a function of x only and y' a function of y only).

The results of Lecornu's efforts are given. The required form becomes

$$Z_1(z)X(x) + Z_2(z)Y(y) = 1$$

which, after partial differentiation and some manipulation, leads to a set of conditions of possibility which are

$$\frac{1}{p}\frac{\partial w}{\partial x} = \frac{1}{q}\frac{\partial w}{\partial y} = u - vw$$

where

$$u = \frac{1}{pq}\frac{\partial^2 \ln q/p}{\partial x \partial y},$$

$$v = \frac{s}{pq},$$

$$w = \left(q\frac{\partial u}{\partial x} - p\frac{\partial u}{\partial y}\right) \Big/ \left(q\frac{\partial v}{\partial x} - p\frac{\partial v}{\partial y}\right),$$

and p, q, and s have their previous meanings.

If the conditions are satisfied, then w is a function of z. A quantity T is calculated from $T = \int w \, dz$.

T is now expressed as a function of x and y using the original relationship $F(x, y, z) = 0$. The relationship (2.45) obtained by Massau; i.e.,

$$\frac{Z_1}{Z_2} = \frac{pY'}{qX'},$$

also applies in Lecornu's proof. Denoting $\ln(Z_1/Z_2)$ by T, $\ln X'$ by $-f$, and $\ln Y'$ by g, the relationship can be expressed in the form

$$f + g = \ln \frac{q}{p} - T.$$

It follows that $\ln(q/p) - T$ is the sum of a function of x and a function of y which are f and g respectively. Then

$$X = \int e^{-f} dx \quad \text{and} \quad Y = \int e^{g} dy$$

are calculated. Finally, the relations $Z_1 X + Z_2 Y = 1$ and $Z_2 = Z_1 e^{T}$ enable Z_1 and Z_2 to be calculated.

It may be instructive to consider the case of $z = x^2 y^2$ to compare the method with that of Massau. Thus,

$$\ln \frac{q}{p} = \ln x - \ln y$$

and

$$\frac{\partial}{\partial x} \ln \frac{q}{p} = \frac{1}{x} \quad \text{and} \quad \frac{\partial^2}{\partial x \partial y} \ln \frac{q}{p} = 0.$$

Therefore $u = 0$, $v = (x^2 y^2)^{-1}$, and $w = 0$ hence

$$u - vw = 0,$$

$$\frac{\partial w}{\partial x} = 0,$$

$$\text{and} \quad \frac{\partial w}{\partial y} = 0$$

and the conditions are satisfied. We now have

$$T = \int 0 \, dz = 0$$

$$\text{and} \quad f + g = \ln x - \ln y$$

giving

$$f = \ln x \quad \text{and} \quad g = -\ln y$$

so that

$$X = \int e^{-\ln x}\, dx = \ln x,$$

$$Y = \int e^{-\ln y}\, dy = \ln y,$$

$$\text{and} \quad Z_1 \ln x + Z_2 \ln y = 1.$$

But $Z_2 = Z_1 e^0$; i.e., $Z_2 = Z_1$, therefore

$$Z \ln(xy) = 1,$$

$$Z = \frac{1}{\ln \sqrt{Z}} = \frac{2}{\ln z},$$

and finally

$$\ln z = 2 \ln x + 2 \ln y.$$

The methods of Massau and Lecornu are similar in the basic philosophy of using partial differentiation; they differ only in detail. Writers on the subject, when they refer to these methods at all, tend only to make the point that Massau's requires four integrations and Lecornu's requires three, but this hardly seems to be a point of any great importance. Both methods are important in that they are successful attacks on an important problem in nomography. Further details of both proofs are given in Appendix A.

It will have been observed that, as in the case of the example which illustrated Saint-Robert's criterion, the examples which illustrate the methods of Massau and Lecornu do not take account of constants of integration. This is because it is a feature of such methods that constants of integration eliminate themselves.

2.7. D'Ocagne's *Nomographie*

D'Ocagne's book *Nomographie – Les Calculs Usuels Effectues au Moyen des Abaques* was published in 1891 [**90**]. The author was twenty-nine years of age and had already a wide practical experience behind him in the Corps des Ponts et Chaussées and on detachment between 1885 and 1889 to the marine hydraulic service. This experience would doubtless have made him aware of the need to perform calculations with speed and with minimum error and this in turn would have heightened his interest in the techniques of geometric computation. Before 1891 no book had been published which dealt with either the principles or general techniques of nomography, and on this score alone the book represents an important event in the development of the subject. The fact that it is a compact, concise and elegant book adds further to its importance; as an introduction to the subject it compares most favorably with books written more than half a century later. It is also the first work which describes nomography by that name.

Earlier sections of this thesis have referred to intermittent papers which have appeared on the subject and we have also seen that in one case, that of Lallemand's hexagonal method, a brochure was published specifically for the use of the Nivellement Général de France and was not made available to the public. We may assume with same confidence that in the period between 1842 and 1891, that is the period between the year in which a law was published committing France to establish a network of railways and the year of the publication of d'Ocagne book, a considerable body of knowledge had been built up within the technical departments of state. In his book, d'Ocagne presents this material and other material known at the time and attempts successfully to extract general principles. much of the material is based on the work of Lallemand and d'Ocagne himself while some is earlier work, but it is all developed consistently and presented in a form which the reader of a more recent text, such as *The Nomogram* [**2**], published in 1963, would instantly recognize. In fact, many types of nomogram are here publicly recorded for the first time.

In his forward, d'Ocagne compares nomography with descriptive geometry and finds a similarity in that the latter rests on the use of a few simple propositions of pure geometry while the former rests on a few principles of analytical geometry. This sets the tone of the book; it is to be a mathematician's book in which underlying principles are all important, rather than a collection of techniques. The importance of the book to a history of nomography is that it represents the state of the art as it was in 1891; no other source gives us such a comprehensive view.

The work begins with equations containing not more than three variables, presenting us with the theory of intersection nomograms. It is presented as follows. If the result of the elimination of x and y from the three equations

$$F_1(x, y, \alpha) = 0, \tag{2.48}$$

$$F_2(x, y, \beta) = 0, \tag{2.49}$$

$$\text{and} \quad F_3(x, y, \gamma) = 0 \tag{2.50}$$

$$\text{is} \quad F(\alpha, \beta, \gamma) = 0 \tag{2.51}$$

then to construct a nomogram of equation (2.51) it is only necessary to construct the three systems of curves defined by equations (2.48), (2.49), and (2.50) in which one varies respectively the parameters α, β, and γ taking care to inscribe the value of the parameter in a suitable manner on each curve. These curves he calls *isoplethés* respectively for the parameters α, β, and γ. Consequently a set of values of the three parameters satisfy equation (2.51) when the corresponding isoplethés meet in a point. Thus, one of the parameters in (2.51) may be found when the two others are given. We recognize here the clarification of an idea from Massau. This then gives the principle behind the simple intersection nomogram of Figure 2.22.

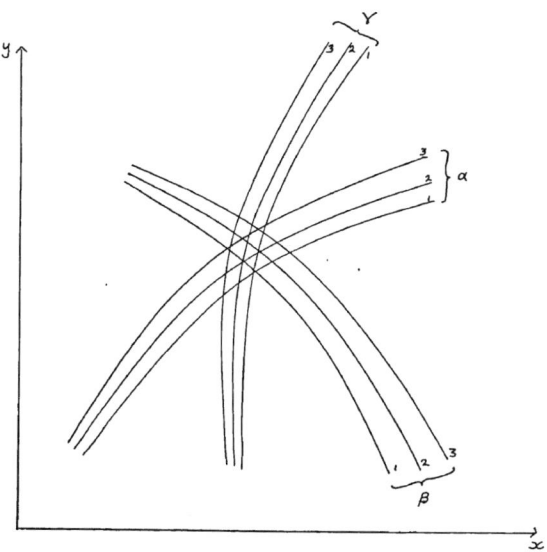

FIGURE 2.22

This idea is now simplified since it is recognized that two of the equations (2.48), (2.49), and (2.50) may be chosen arbitrarily. If, for example, we so choose equations (2.48) and (2.49) then equation (2.50) can be obtained by eliminating α and β between them and the given equation (2.51). It makes practical sense to choose equations (2.48) and (2.49) to be as simple as possible. We therefore put $x = \alpha$ and $y = \beta$ when equation (2.50) becomes $F(x, y, \gamma) = 0$. We thus have a nomogram of the type shown in Figure 2.23.

We recognize here that we have arrived back at a chart of a type very like that of Pouchet for $z_1 z_2 = z_3$, given in Figure 1.1, but the difference here is that d'Ocagne has given the principle on which it is based.

The natural topic to continue with is anamorphosis and this is what d'Ocagne does starting with a simple illustration which is no more than a tidier version of Lalanne's earlier work. Taking as his example an equation which is a variant of that suggested to Lalanne by Cauchy,

$$F(\alpha, \beta, \gamma) = f(\alpha)\psi_1(\gamma) + \phi(\beta)\psi_2(\gamma) + \psi_3(\gamma) = 0, \qquad (2.52)$$

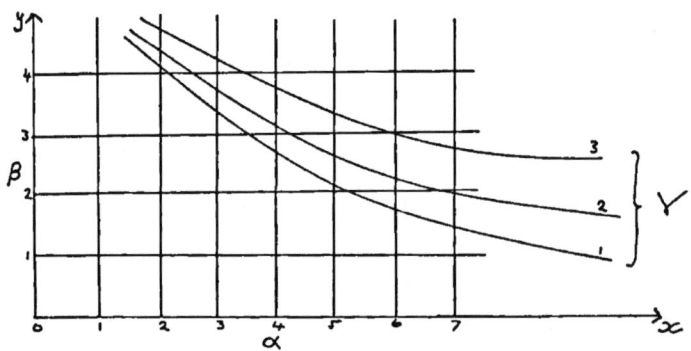

FIGURE 2.23

he takes for his first two equations

$$x = f(\alpha) \tag{2.53}$$

$$\text{and} \quad y = \phi(\beta) \tag{2.54}$$

whence he arrives at

$$x\psi_1(\gamma) + y\psi_2(\gamma) + \psi_3(\gamma) = 0. \tag{2.55}$$

Thus the isoplethés for α and β are again parallel to the axes but this time not equally spaced.

The isoplethés for γ are also straight lines which are tangents to a curve which can be found, if it is wanted, by eliminating γ between equation (2.55) and its derivative with respect to γ.

This is also Lalanne's work put in a more mathematical form and made more concise. It is worth noting that the gap between Lalanne's paper on anamorphosis and the publication of those same ideas, concisely and mathematically expressed, is forty-five years.

D'Ocagne next examines an idea that had been expressed seven years earlier by Massau. He considers the problem of looking for the general form of equations which are representable by three streams of linear isoplethés.

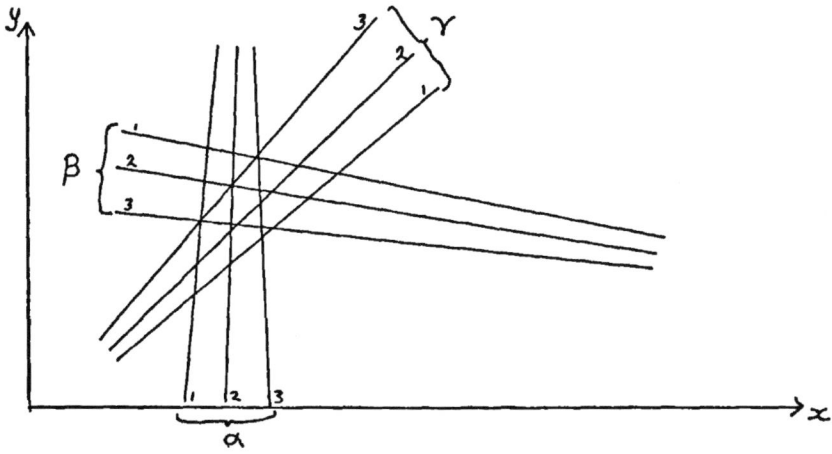

FIGURE 2.24

He observes that $F_1(x, y, \alpha)$, $F_2(x, y, \beta)$, and $F_3(x, y, \gamma)$ must have the following forms:

$$F_1 = xf_1(\alpha) + yf_2(\alpha) + f_3(\alpha) = 0,$$
$$F_2 = x\phi_1(\beta) + y\phi_2(\beta) + \phi_3(\beta) = 0, \qquad (2.56)$$
$$\text{and} \quad F_3 = x\psi_1(\gamma) + y\psi_2(\gamma) + \psi_3(\gamma) = 0$$

and that the form of $F(\alpha, \beta, \gamma) = 0$ must therefore be

$$\begin{vmatrix} f_1(\alpha) & f_2(\alpha) & f_3(\alpha) \\ \phi_1(\beta) & \phi_2(\beta) & \phi_3(\beta) \\ \psi_1(\gamma) & \psi_2(\gamma) & \psi_3(\gamma) \end{vmatrix} = 0 \qquad (2.57)$$

when these conditions are satisfied. Nomograms of the type shown in Figure 2.24 are then obtained.

D'Ocagne then makes the remark, which in the light of subsequent work must be regarded as a considerable understatement, that "It is not always easy to see whether a given equation in three variables can be put into this form." He adds a footnote which is worth quoting in full.

"The common character of all equations susceptible of revert-
ing to the determinant form (above) express themselves by partial differential equations obtained by the elimination of the arbitrary functions which enter into that form. Those functions
are six in number (because on each line of the determinant
one function must be a linear combination of the other two),
and the analytic problem consisting of eliminating them will
not want of a certain complication. This problem has been
completely resolved, and in a very elegant manner by M. Ing
des Mines Lecornu (C.R. tCII p. 815) in the case where the
determinant form reduces to

$$f(\alpha)\psi_1(\gamma) + \phi(\beta)\psi_2(\gamma) + \psi_3(\gamma) = 0.$$

M. Lecornu has not only eliminated the four arbitrary functions
that limit (the above) but also shows the way in which they can
be determined when one has verified that the form is possible."

We observe that no mention is made of the work of Massau nor of that of
Saint-Robert. It is possible that d'Ocagne was unaware of both of these efforts
in 1891. However, when his *Traité de Nomographie* appeared in 1899, d'Ocagne
devoted some space to both [**92**].

He gives a case, frequent enough in practice he claims, where it is easy to
verify that the equation can be put into the determinant form, namely

$$\chi_1(\alpha,\beta)\psi_1(\gamma) + \chi_2(\alpha,\beta)\psi_2(\gamma) + \chi_3(\alpha,\beta)\psi_3(\gamma) = 0 \qquad (2.58)$$

in which it is sufficient to put

$$x = \frac{\chi_1(\alpha,\beta)}{\chi_3(\alpha,\beta)} \quad \text{and} \quad y = \frac{\chi_2(\alpha,\beta)}{\chi_3(\alpha,\beta)}$$

and to eliminate, in turn, and β and α from these equations. If the result of
the eliminations are of the first degree in x and y; i.e.,

$$x f_1(\alpha) + y f_2(\alpha) + f_3(\alpha) = 0 \qquad (2.59)$$
$$\text{and} \quad x\phi_1(\beta) + y\phi_2(\beta) + \phi_3(\beta) = 0; \qquad (2.60)$$

then these equations are used. A third equation is obtained by eliminating α
and β between (2.59), (2.60) and (2.58). Since system (2.59) and system (2.60)

are equivalent to the system

$$x = \frac{\chi_1(\alpha, \beta)}{\chi_3(\alpha, \beta)} \quad \text{and} \quad y = \frac{\chi_2(\alpha, \beta)}{\chi_3(\alpha, \beta)}$$

the third equation will be

$$x\psi_1(\gamma) + y\psi_2(\gamma) + \psi_3(\gamma) = 0.$$

D'Ocagne considers it desirable to give a name to an equation which is representable by three sets of linear isoplethés and proposes the term *equations à triple reglure*.

In passing the remark is made, again echoing an idea of Lalanne, that circles are almost as easy to draw as straight lines and that therefore, for certain cases which are not à triple reglure, circular isoplethés may be the answer.

Binary scales, due to Lallemand, are briefly discussed. The idea of such a scale is that it is constructed so that one point may have two interpretations. They are not of great importance to this study. Examples appear in Lallemand's nomogram of Figure 2.21.

A second idea attributed to Lallemand is also briefly described. It is that of the graphical elimination of a variable between two equations. For example consider

$$F(\alpha, \beta, \gamma) = 0$$
$$\text{and} \quad \phi(\alpha, \beta, \gamma) = 0.$$

One can construct both nomograms taking in each the scale $y = \gamma$ as shown in Figure 2.25.

Given for example $\alpha = 2$, $\beta = 1$, and $\beta' = 3$ it is easy to see that $\alpha' = 3$ by following the broken line.

Although γ is 5 in this case the value plays a passive role. It can be seen that the method could be extended by allowing the scales of α and α' to coincide, if this was thought desirable. We note that this idea of graphical elimination had also been given by Massau in 1884.

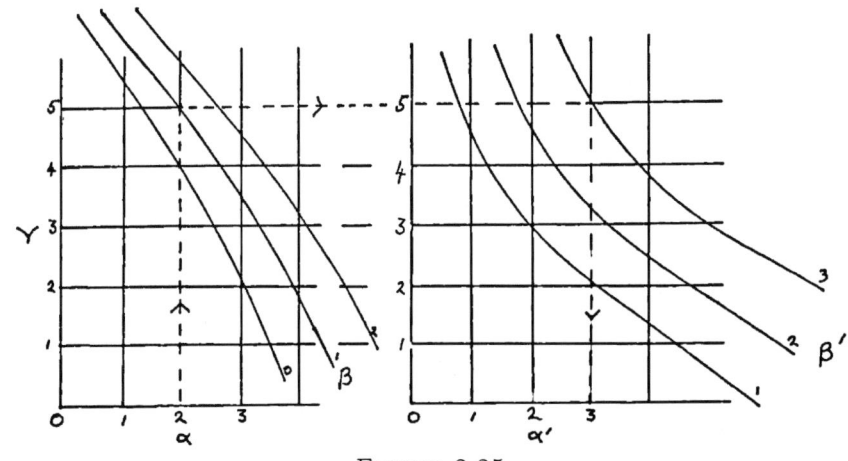

FIGURE 2.25

The remainder of the book is devoted mainly to the elaboration of ideas already described in this thesis. A large section deals with Lallemand's hexagonal nomogram and in particular the extension of it to deal with equations having more than three variables. Graphical addition and graphical multiplication are dealt with in full, but these are forms of the hexagonal nomogram and, although ingenious, add little to the advancement of the subject.

As one would have expected, d'Ocagne's earlier work on parallel coordinates and the related nomograms is dealt with at some length. Here, however, unlike in his 1884 paper, he uses the term *points isoplethés* to describe the duals of his isopleth lines. Otherwise the work in this area is repeated with more elaborate examples.

Finally, one must note d'Ocagne's brief reference to the principle of homography. This is important because he is suggesting the use of geometric principles to change the appearance of a nomogram so that it may be read with greater ease. He merely notes that the transformations

$$x' = \frac{a_0 x + b_0 y + c_0}{dx + ey + f}$$

$$\text{and} \quad y' = \frac{a_1 x + b_1 y + c_1}{dx + ey + f}$$

transform points on a straight line to other points on a straight line. However, homographic, or projective, transformations have become most important features of nomography.

We thus have a clear picture of the state of nomography in 1891. Intersection nomograms were well established; alignment nomograms less so but they were on a sound theoretical basis; hexagonal nomograms had been developed to their full extent; anamorphosis was being regarded as an important problem, although perhaps being underestimated, and the idea of using geometric projections to improve nomograms had been sown.

2.8. The State of Nomography in 1893

Two events took place in 1893 which can be taken as indicators of, on the one hand, the status of d'Ocagne and his new discipline, and on the other, of the attention that British engineers had paid to nomography.

The first event was the International Mathematical Congress held in connection with the World's Columbian Exposition in Chicago. Many famous mathematicians read papers at this conference. As examples we can cite Charles Hermite, who read a paper on elliptical functions and David Hilbert who read one on the theory of invariants. Also reading a paper was d'Ocagne. His paper was called "Nomography: On equations representable by three linear systems of isoplethé points" [91].

If one can judge a man by the company that he keeps then it seems that by 1893 d'Ocagne's standing as a mathematician was high. One must assume that some kind of selection or invitation was necessary before a paper could be presented at such a conference. Of course, d'Ocagne does seem to have been something of a showman, he had more than a passing connection with the theater, and he did not let an opportunity pass that would enable him to publicize nomography, but I think it must be accepted that by this time he was an important mathematician.

The paper itself is interesting and well presented. It deals with a special aspect of alignment nomograms, those in which the three scales, or systems of isoplethé points as he likes to call them, are on straight lines and in particular when the points on those straight lines are equally spaced. It is a paper ahead of its time on a theme which d'Ocagne would return to four years later. The other event of 1893 was a report to the British Association for the Advancement of Science on "Graphical Methods in Mechanical Science." This was the third part of a report, the preliminary part of which had been presented in 1889 with a second part in 1892. The author was Professor H.S. Hele-Shaw of University College, Liverpool.

It is no surprise that these reports deal largely with graphical statics, but it is a little surprising that some recent nomographic ideas have no place in them. The nearest that one comes to nomography is in the 1892 report where the following appears:

> "Falling under the head of 'graphical tables' are the constructions devised by M. L. Lalanne. [...] An example of one of these tables, called by the inventor an 'abacus', was shown as a wall diagram to the Mechanical Section. The ordinates and abscissae of this diagram are not numbered according to their actual values, but are logarithmic, exactly as the scale on a slide rule. By means of this diagram operations of multiplication can at once be performed, and by a slight modification products such as a^2b and the $\sqrt{ab}$ can be readily obtained."

The ideas of nomography had therefore not penetrated to the British engineering establishment by 1892 except for an idea from 1846. This may have been due to poor communication of ideas; it may also have been due to the poor mathematical background of many British engineers, for elsewhere the report suggests that graphical statics are preferred to calculations for that very reason.

However, British military engineers were a little more forward looking, for also in 1893 there appeared, in the Professional Papers of the Corps of Royal Engineers, an article on the "Graphic Solution for Equations of the second, third and fourth powers" which was truly nomographic in spirit [17]. It was a translation by Major W.H. Chippindall, R.E. of a paper by Lt. Julius Mandl of the Imperial Austrian Engineers. The work has a very practical purpose as can be seen from the opening paragraph:

> "In order to avoid the use of logarithmic tables and the necessity of obtaining the second and third roots in the solution of equations of the third and fourth power, the accompanying table was constructed for solutions in which greater exactness was not required than the first two or three figures."

The accompanying table referred to is a graphical table making use of a transparency which itself can be constructed by tracing lines already on the table. The theory is based on the fact that various sums and products of the roots of the equations may be expressed in terms of the coefficients of those equations.

In the case of the second degree equation,

$$x^2 + Ax + B = 0, \tag{2.61}$$

we have, if we suppose the roots to be x_1 and x_2,

$$x_1 + x_2 = -A \tag{2.62}$$

$$\text{and} \quad x_1 x_2 = B. \tag{2.63}$$

Equation (2.62) represents a straight line, supposing x_1 and x_2 to be Cartesian coordinates. Furthermore, whatever the value of A, the straight line always has a slope of -1.

Equation (2.63) represents, for varying B, a set of rectangular hyperbolæ referred to their common asymptotes as axes.

The intersection of (2.61) with (2.63) will then have for abscissa and ordinate respectively the values of x_1 and x_2 which satisfy equation (2.61) for some given A and B. The proposal is therefore to construct such hyperbolæ on a rectangular Cartesian framework and to have a straight line on a transparency which can be used to represent a line of slope -1 for any A, enabling the appropriate values of x_1 and x_2 to be read. If the whole diagram is square then the line with a slope of -1 is parallel to the diagonal joining top left to bottom right. Figure 2.26, which is not to scale, shows the general appearance.

Moving to equations of the third degree,

$$x^3 + Ax^2 + Bx + C = 0, \tag{2.64}$$

if we take the roots to be x_1, x_2, and x_3 then

$$x_1 + x_2 + x_3 = -A, \tag{2.65}$$

$$x_1 x_2 + x_1 x_3 + x_2 x_3 = B, \tag{2.66}$$

$$\text{and} \quad x_1 x_2 x_3 = -C. \tag{2.67}$$

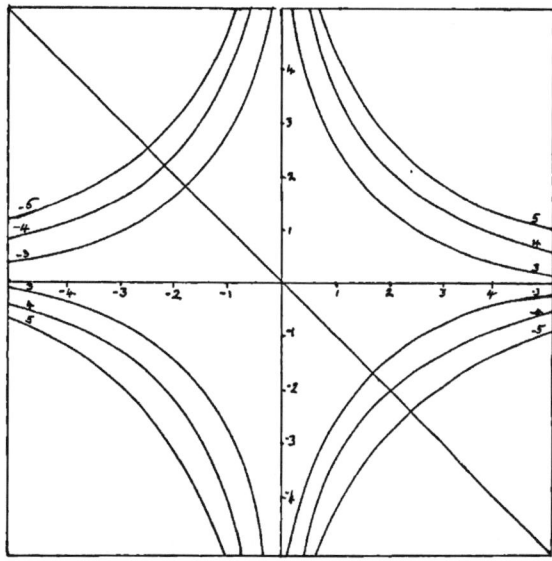

$$\textsc{Figure } 2.26$$

Letting $x_2 + x_3 = z$ and $x_2 x_3 = y$, then

$$x_1 + z = -A, \tag{2.68}$$

$$x_1 z + y = B, \tag{2.69}$$

$$\text{and} \quad x_1 y = -C. \tag{2.70}$$

If we eliminate z between (2.68) and (2.69), then we have

$$y = x_1^2 + A x_1 + B. \tag{2.71}$$

If we now consider x_1 and y to be co-ordinate axes with x_1 as the abscissa and y as the ordinate, then (2.70) again represents a rectangular hyperbola while (2.71) is a parabola with the constant equal to unity, the axis parallel to the ordinate axis and the vertex always downwards.

Equation (2.61) may be rewritten

$$y - \left(B - \frac{A^2}{4} \right) = \left(x_1 + \frac{A}{2} \right)^2 \tag{2.72}$$

showing that the vertex is given by

$$\left(\frac{-A}{2}, B - \frac{A^2}{4}\right).$$

It follows that a knowledge of A and B, which we have, will fix the position of the vertex but otherwise all parabolae given by (2.71) are the same shape. Therefore our transparency should also carry this parabola, its axis and the tangent to its vertex. The axis can be the straight line referred to earlier. The intersection of the parabola with the hyperbola is equivalent to the simultaneous solution of (2.70) and (2.71); that is, to the solution of

$$x_1^3 + Ax_1^2 + Bx_1 + C = 0.$$

It follows that to solve (2.64) we must find the co-ordinates of the vertex of the parabola; position the transparency so that its parabola has its vertex at the point with those coordinates and its axis parallel to the ordinate axis, and then note the abscissa of the points of intersection of the parabola with the hyperbola given by (2.70). Normally there will be three such abscissae corresponding to the roots of (2.64), x_1, x_2, and x_3.

If the scale is such that only two intersections are given; i.e., we only have x_1 and x_2, the third root will be given by (2.65); i.e., $x_3 = -A - x_1 - x_2$. If the equation is such that the intersections fall outside the diagram, then substitutions of the type $x = m\xi$ or $x = \xi + m$ will correct the situation. For example,

$$x^3 - 38x^2 + 461x - 1768 = 0$$

becomes

$$\xi^3 - 8\xi^2 + \xi + 42 = 0$$

after the substitution $x = \xi + 10$ has been made, the roots of the original equation being given by $\xi_1 + 10$, $\xi_2 + 10$, and $\xi_3 + 10$.

For the equation of the fourth degree

$$x^4 + Ax^3 + Bx^2 + Cx + D = 0 \qquad\qquad (2.73)$$

we have

$$x_1 + x_2 + x_3 + x_4 = -A, \tag{2.74}$$

$$x_1x_2 + x_1x_3 + x_1x_4 + x_2x_3 + x_2x_4 + x_3x_4 = B, \tag{2.75}$$

$$x_1x_2x_3 + x_1x_2x_4 + x_1x_3x_4 + x_2x_3x_4 = -C, \tag{2.76}$$

$$\text{and} \quad x_1x_2x_3x_4 = D. \tag{2.77}$$

The substitutions

$$x_1 + x_2 = m, \quad x_3 + x_4 = n, \quad x_1x_2 = p, \quad \text{and} \quad x_3x_4 = q$$

reduce equations (2.74), (2.75), (2.76), and (2.77) to

$$m + n = -A, \tag{2.78}$$

$$mn + p + q = B, \tag{2.79}$$

$$mq + np = -C, \tag{2.80}$$

$$\text{and} \quad pq = D. \tag{2.81}$$

Equations (2.78) and (2.80) give

$$m = -\frac{Ap - C}{p - q} \quad \text{and} \quad n = \frac{Aq - C}{p - q}.$$

Substituting in (2.79), and using (2.81) with the substitution $p + q = z$, we get

$$z^3 - Bz^2 + (AC - 4D)z - [C^2 + D(A^2 - 4B)] = 0 \tag{2.82}$$

Thus the fourth degree equation (2.73) can be transformed into the third degree equation (2.82), for which we can obtain a root by the method already described.

Having obtained the root z, we can find p and q from $p + q = z$ and $pq = D$ and then m and n from $m + n = -A$ and $mn = B - z$.

Thus, m, n, p, and q are all that are required to obtain x_1, x_2, x_3, and x_4. Although (2.82) may have three roots it is immaterial which of the three are used as all lead to the same set of values for x_1, x_2, x_3, and x_4. Figure 2.27 is reproduced from Major Chippindall's original paper. P is the parabola which must be transferred to the transparency as also must its axis ab, the tangent to its vertex, CD, and the line AB which contains the point a.

As has been noted, Major Chippindall's paper, or perhaps we should say Lt. Mandl's, is nomographic in spirit in that solutions are provided for a whole class of equations in which the coefficients are considered to be the variables. There is, of course, no reason to believe that the ideas owe anything to the work of Lalanne, Massau or d'Ocagne. What is of interest is the reason why an army engineer should appear to have a better grasp of the importance of geometric computation than the civilian engineers who reported to the British Association. The reason is that officers of the Royal Engineers at this time were much better educated in science and mathematics than the civilian engineers.

During the latter part of the nineteenth century there was such concern in Britain over the state of technical education that, in March 1863, the government ordered a Parliamentary Select Committee, chaired by Bernhard Samuelson, to look into the problem. In its report the committee said that "a hindrance second only to the defective elementary education of the pupils is the scarcity of science teachers and the want of schools for training them."

Although the reasons may be different, the situation one hundred years later seems to have a certain similarity. One of the measures taken by the government to remedy this situation was to allow officers of the Royal Engineers to supervise the examinations of the Department of Science and Art and to inspect science teaching in schools. The reason given for this solution was that nowhere else could a suitable body of science inspectors be found.

It only became possible to dispense with their services towards the end of the century. The report of the Board of Education for 1913 – 1914, stated that officers of the Royal Engineers "in the early days of the Department were one of the few bodies of men in the country with an organized scientific training [15]." Later, we shall see that d'Ocagne's ideas were also introduced into Britain by army officers, this time of both the Royal Engineers and the Royal Artillery.

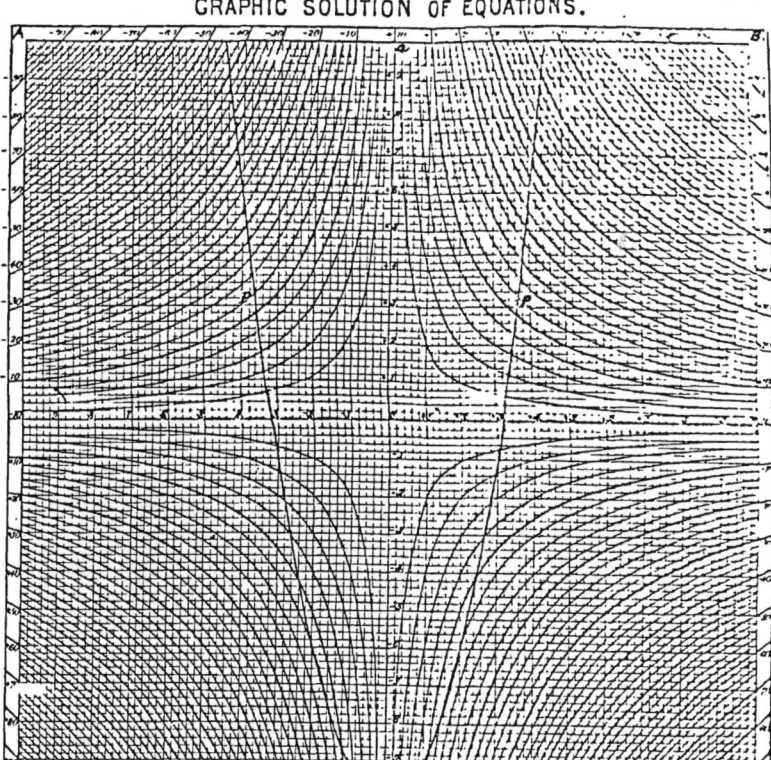

FIGURE 2.27. The chart which accompanied Major Chippindall's paper

2.9. A Theoretical Problem of Alignment Nomograms

So far theoretical problems in nomography had been concerned with anamorphosis and as such had been related to intersection nomograms. However, the alignment nomogram developed by d'Ocagne has a parallel theoretical problem. Briefly, it is this. Suppose the three curves of Figure 2.28 to be given by the parametric equations,

$$x : \xi = \frac{\phi_1(x)}{\phi_3(x)}, \qquad \eta = \frac{\phi_2(x)}{\phi_3(x)};$$

$$y : \xi = \frac{\psi_1(y)}{\psi_3(y)}, \qquad \eta = \frac{\psi_2(y)}{\psi_3(y)};$$

$$z : \xi = \frac{\theta_1(z)}{\theta_3(z)}, \qquad \eta = \frac{\theta_2(z)}{\theta_3(z)}.$$

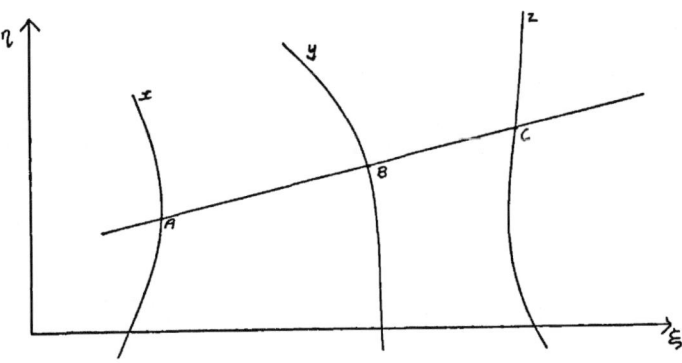

FIGURE 2.28

If the straight line AC cuts the curves as shown then the values of x at A, of y at B and of z at C must be connected by the relation

$$\left(\frac{\theta_2(z)}{\theta_3(z)} - \frac{\psi_2(y)}{\psi_3(y)} \right) \bigg/ \left(\frac{\theta_1(z)}{\theta_3(z)} - \frac{\psi_1(y)}{\psi_3(y)} \right) = \left(\frac{\psi_2(y)}{\psi_3(y)} - \frac{\phi_2(x)}{\phi_3(x)} \right) \bigg/ \left(\frac{\psi_1(y)}{\psi_3(y)} - \frac{\phi_1(x)}{\phi_3(x)} \right)$$

since the gradient of AB is the same as that of BC. This relation can be expressed in determinant for as

$$\begin{vmatrix} \phi_1(x) & \phi_2(x) & \phi_3(x) \\ \psi_1(y) & \psi_2(y) & \psi_3(y) \\ \theta_1(z) & \theta_2(z) & \theta_3(z) \end{vmatrix} = 0.$$

The theoretical problem is the following:

Given a relationship $F(x, y, z) = 0$, can it be expressed in the determinant form given?

If it can, then an alignment nomogram can be constructed. This theoretical problem of alignment is the same problem as that revealed in connection with anamorphosis by d'Ocagne, (2.56) and (2.57). The same, that is, from a pure mathematical point of view; the geometrical interpretation differs according to the particular case. I have found no evidence to suggest that the dual nature of the problem had been appreciated by the end of the nineteenth century.

The first attack on the alignment problem was by Ernest Duporcq in 1898 [**31**]. He gave seven conditions which, if satisfied, showed that such a form was possible and also provided enough information to enable that form to be obtained. However, the method is more satisfactory at the theoretical level than at the practical level, it being very cumbersome for a moderately complex problem. Yet, as the first attack, it is one that doubtless had influence on later attempts. The authors of these make reference to Duporcq and it is therefore of some interest to indicate his approach, particularly since his paper is rather obscure.

Although not said explicitly, Duporcq's starting point is the recognition that if $F(x, y, z) = 0$ can be expressed in the determinant form then it must be capable of being written as,

$$F(x, y, z) = P_1(x)R_1(y, z) + P_2(x)R_2(y, z) + P_3(x)R_3(y, z) = 0. \qquad (2.83)$$

With this in mind he is able to construct three 4×4 determinants containing $F(x, y, z)$ and forms of it involving three arbitrary and distinct values of x, namely a, a', and a''; and similarly of y, viz. b, b', and b''; and of z, viz. c, c', and c'' so that the determinants are all identically zero. One of them is given here:

$$
\begin{vmatrix}
F(x,y,z) & F(x,b,c) & F(x,b',c') & F(x,b'',c'') \\
F(a,y,z) & F(a,b,c) & F(a,b',c') & F(a,b'',c'') \\
F(a',y,z) & F(a',b,c) & F(a',b',c') & F(a',b'',c'') \\
F(a'',y,z) & F(a'',b,c) & F(a'',b',c') & F(a'',b'',c'')
\end{vmatrix} . \tag{2.84}
$$

To demonstrate that (2.84) is identically zero it is only necessary to consider each function F replaced by the corresponding expression given by (2.83). The determinant (2.84) is then seen to be the sum of 81 determinants of which the following is an example:

$$
R_1(y,z)R_2(b,c)R_3(b',c')R_1(b'',c'')
\begin{vmatrix}
P_1(x) & P_2(x) & P_3(x) & P_1(x) \\
P_1(a) & P_2(a) & P_3(a) & P_1(a) \\
P_1(a') & P_2(a') & P_3(a') & P_1(a') \\
P_1(a'') & P_2(a'') & P_3(a'') & P_1(a'')
\end{vmatrix} .
$$

This determinant is identically zero since the first and last columns are the same. The other 80 determinants, by virtue of the way in which they have been constructed, must each contain at least two identical columns and consequently must each be identically zero. It follows that the composite determinant is identically zero.

The knowledge that (2.84) is identically zero leads, on expanding along the first rows, to the following expression:

$$
\begin{aligned}
F(x,y,z) = {} & U(y,z)F(x,b,c) + \\
& V(y,z)F(x,b',c') + \\
& W(y,z)F(x,b'',c'').
\end{aligned} \tag{2.85}
$$

So far we have (2.84) written in the form of (2.83) but with only the functions in x known.

It should be noted that (2.83) is one of three possible forms that must hold if the problem is to be resolved, the others would have isolated the variable y in one case and z in the other in the same way that (2.83) has isolated x. To each of these two forms will correspond an appropriate determinant similar to, but different from, (2.84). It is for this reason that Duporcq has imposed three determinant conditions on a, a', a'', b, b', b'', and c, c', c''. Equation (2.85) enables three functions $U(y, z)$, $V(y, z)$, and $W(y, z)$ to be found. It is next necessary to isolate the separate functions of y and z. Let them be $g_1(y)$, $g_2(y)$, $g_3(y)$, $h_1(z)$, $h_2(z)$, and $h_3(z)$.

Then,

$$F(x, y, z) = \begin{vmatrix} F(x, b, c) & F(x, b', c') & F(x, b'', c'') \\ m_1 g_1(y) & m_2 g_2(y) & m_3 g_3(y) \\ n_1 h_1(z) & n_2 h_2(z) & n_3 h_3(z) \end{vmatrix}$$

where m_1, m_2, m_3, n_1, n_2, and n_3 are constants. Therefore it is required that

$$U(y, z) = m_2 n_2 g_2(y) h_3(z) - m_3 n_2 g_3(y) h_2(z)$$
$$= \lambda_1 A_1(y, z) - \mu_1 B_1(y, z);$$

i.e., the determinant

$$\begin{vmatrix} U(y, z) & A_1(y, z) & B_1(y, z) \\ U(b, c) & A_1(b, c) & B_1(b, c) \\ U(b', c') & A_1(b', c') & B_1(b', c') \end{vmatrix}$$

must be identically zero.

A similar condition applies for the determinants in V, A_2, B_2 and W, A_3, B_3. These are then three more conditions to be satisfied. From them can be obtained λ_1 and μ_1 as illustrated above, and similarly λ_2, μ_2, λ_3, and μ_3. The values of the λ's and μ's are known in terms of the m's and n's. They are

$$\lambda_1 = m_2 n_3, \quad \mu_1 = m_3 n_2,$$
$$\lambda_2 = m_3 n_1, \quad \mu_2 = m_1 n_3,$$
$$\lambda_3 = m_1 n_2, \quad \mu_3 = m_2 n_1$$

from which the seventh condition follows, that

$$\lambda_1 \lambda_2 \lambda_3 = \mu_1 \mu_2 \mu_3.$$

One problem still remains and that is to find the A's and B's. Since they are combinations of g's and h's the problem is reduced to finding these.

Consider the relationships,

$$V(y, z) = \lambda_2 g_3(y) h_1(z) - \mu_2 g_1(y) h_3(z)$$
$$\text{and} \quad W(y, z) = \lambda_3 g_1(y) h_2(z) - \mu_3 g_2(y) h_1(z).$$

Putting in turn $z = c$ and $z = c'$ in the first relationship and combining the two resulting expressions we get

$$g_1(y) = \beta_1 V(y, c) + \beta_2 V(y, c').$$

Treating the second one similarly we get an alternate expression

$$g_1(y) = \gamma_1 W(y, c) + \gamma_2 W(y, c')$$

where β_1, β_2, γ_1, and γ_2 are constants. The other values can be found in a similar fashion.

The importance of Duporcq's paper lies in the fact that it focused attention on an important problem and suggested a line of approach for its solution. It has little value as a practical aid.

2.10. The Lives of Lalanne, Lallemand and d'Ocagne

Many of those who have been concerned with nomography have been very interesting men. The lives of three of the main characters in the early development of the subject are briefly described here.

2.10.1. Léon Lalanne. Lalanne was born in Paris in 1811, became a student first at l'École Polytechnique in 1829 and then at l'École des Ponts et Chaussées in 1831. In 1846 he was a responsible engineer engaged on the construction of railways from Paris. After the revolution of 1848 he became Commandant of the 11th legion of the National Guard. In 1849 he was arrested but released almost immediately and, following a coup d'état, he lived abroad for many years being engaged on public works in, amongst other places, Spain and Switzerland. He returned to France in 1850 and from 1877 until his retirement he was the director of l'École des Ponts et Chaussées. Later he was President of the board of the Omnibus Company of Paris. He was responsible for perfecting many calculating devices and was the author or part author of many publications, including works on the Paris Metro. He was elected to the Academy of Science in 1879, and he died in 1892.

2.10.2. Charles Lallemand. Fewer details of the life of Lallemand have been recorded than of the lives of Lalanne and d'Ocagne. He was born in Meuse during 1657 and died in Haute Marne during 1938. Amongst the appointments which he held were Inspector General of Mines and Director of the Nivellement Général do la France. He was elected to the Academy of Science in 1910.

2.10.3. Maurice d'Ocagne. D'Ocagne was born in Paris in 1852. He was an engineer of the Corps des Ponts et Chaussées but from 1805 to 1889 he was seconded to the naval hydraulic services, first at Rochefort and then at Cherbourg. He became director of maps and charts at the Nivellement Général in 1901 and Inspector General of roads and bridges in 1920. He devoted much time to the teaching of geometry, presumably as a part-time teacher; he was appointed Professor at l'École des Ponts et Chaussées in 1894 and at l'École Polytechnique in 1912. In addition to his expertise in nomography, he was an authority on calculating machines and a student of the history of mathematics.

There was another side to his abilities; under the pseudonym of Pierre Delix he ventured into writing. The most notable of his literary efforts was a one act comedy, called *La Candidate*, which was given more than one hundred performances at the Cluny Theater in Paris during 1888 and 1889. He was elected to the Academy of Science in 1922 and died at La Havre in 1938.

2.11. The End of the Century

The year 1899 saw the publication of d'Ocagne's large work *Traité de Nomographie* [**92**]. This work gave theory and practical examples and was the first standard text on the subject. With its publication, nomography could be regarded as a distinct discipline for it now had all the hallmarks; a set of general principles, the theoretical problems of anamorphosis and alignment and now a treatise.

CHAPTER 3

Development in the Early Part of the 20th Century

A short span of years extending either side of 1900 was very fruitful in the production of ideas which helped nomography to establish itself as a branch of mathematics. The theoretical problem which Duporcq attacked in 1898 has already been noted; namely, given a relationship $F(x, y, z) = 0$, can it be expressed in the determinant form

$$\begin{vmatrix} \phi_1(x) & \phi_2(x) & \phi_3(x) \\ \psi_1(y) & \psi_2(y) & \psi_3(y) \\ \theta_1(z) & \theta_2(z) & \theta_3(z) \end{vmatrix} = 0?$$

If it can, then an alignment nomogram is possible.

However, this problem is the ultimate one in nomography and there are others, more easy to deal with, which were being considered during this period. A line of thought was being explored which related the algebra of a relationship to the geometry of a corresponding nomogram. For example, in the case of the relationship $F(x, y, z) = 0$, an alignment nomogram could have a straight line as a carrier for each of the variables. A term often used for such a carrier is *support*, a term which I will also use when it is helpful to do so. It is also possible that $F(x, y, z) = 0$ could lead to a nomogram having three curved supports or to some intermediate combination of straight lines and curves. Again, the supports could be concurrent or nonconcurrent. All of these properties may be revealed by an examination of the relationship $F(x, y, z) = 0$ and it was to this type of problem that minds turned at this time.

D'Ocagne was the first to investigate this problem. The paper which he presented at the Chicago exposition of 1893 was on alignment nomograms having three rectilinear supports. The subject was given a more rigorous treatment in 1897 when he published a paper in *Acta Mathematica*; a paper which was incorporated into his *Traité de Nomographie* two years later [**92**]. This will be examined shortly when the *Traité* is reviewed.

In 1901, Rodolphe Soreau proposed a classification system based on the linear dependance of the component functions of $F(x, y, z) = 0$. Putting this expression in determinant form, for in this aspect of the work the assumption is made that this is possible, the following compact form is obtained:

$$\begin{vmatrix} f_1(\alpha_1) & g_1(\alpha_1) & 1 \\ f_2(\alpha_2) & g_2(\alpha_2) & 1 \\ f_3(\alpha_3) & g_3(\alpha_3) & 1 \end{vmatrix} = 0.$$

The supports in the xy plane are given by the equations,

$$x = f_1(\alpha_1), \quad x = f_2(\alpha_2), \quad x = f_3(\alpha_3)$$
$$\text{and} \quad y = g_1(\alpha_1), \quad y = g_2(\alpha_2), \quad y = g_3(\alpha_3).$$

Consider the support of α_1. If a linear relationship exists between $f_1(\alpha_1)$ and $g_1(\alpha_1)$; i.e.,

$$f_1(\alpha_1) = c_0 + c_1 g_1(\alpha_1)$$

in which $(c_0^2 + c_1^2) \neq 0$, then $f_1(\alpha_1)$ and $g_1(\alpha_1)$ are said to be *linear dependent*. It follows from this linear dependence that a linear relationship exists between x and y, in fact it is $x = c_0 + c_1 y$. Therefore the support of α_1 will be a straight line. On the other hand, if no such linear relationship exists; i.e., if the only possible values for the c's are $c_0 = c_1 = 0$, then the support given by $x = f_1(\alpha_1)$ and $y = g_1(\alpha_1)$ will be some curve other than a straight line. Thus, the number of linearly independent functions in $F(x, y, z)$ determines the nature of the supports. This number Soreau called the *nomographic order* of $F(x, y, z) = 0$ [**126**].

An examination of the determinant form

$$\begin{vmatrix} f_1(\alpha_1) & g_1(\alpha_1) & 1 \\ f_2(\alpha_2) & g_2(\alpha_2) & 1 \\ f_3(\alpha_3) & g_3(\alpha_3) & 1 \end{vmatrix} = 0$$

shows that in this case the nomographic order is one of 3, 4, 5, or 6. If it is 3, then there are three rectilinear supports; if it is 4, then there are two rectilinear and one curved supports; if it is 5, then there are one rectilinear and two curved supports and if it is 6, then there are three curved supports.

During the following year d'Ocagne introduced his own classification system [**94**]. This is based on the geometry of a nomogram; one with three rectilinear supports is of genus 0, two rectilinear and one curved supports give a nomogram of genus 1 and so on. In fact the genus of a nomogram is the order of that nomogram less three. It is most unlikely that d'Ocagne was unaware of Soreau's system and if this is the case then the introduction of his own system looks rather like an attempt to keep himself to the fore of developments and is not to his credit. Certainly, Soreau felt this and it rankled him for years. In 1922 he asked why d'Ocagne had "called nomographic genus my nomographic order," pointing out that it was only a question of scaling down the number [**128**].

3.1. D'Ocagne's *Traité de Nomographie*

D'Ocagne published his *Traité de Nomographie* [**92**] in 1899. The introduction is dated May 15, 1899, and one may assume that the twentieth century had dawned by the time the work was in general circulation. As far as I am aware the work has never been translated from the French and yet this in no way seems to have diminished its influence. The evidence of personal acknowledgement or some less direct evidence, such as an unusual use of symbols, suggests that d'Ocagne's ideas spread rapidly to America, Britain, the rest of the continent of Europe and Russia.

In the main, the work is a complete summary end description of existing knowledge illustrated with many examples. The chapter headings show this; they are

    I. Equations with Two Variables
    II. Equations with Three Variables - Intersecting Nomograms
    III. Equations with Three Variables - Alignment Nomograms
    IV. Systems of Two Equations
    V. Equations with More than Three Variables
    VI. General Theory. Analytical Developments

Chapters IV and V, while based on earlier work, also contain certain technical additions which are little more than methods for linking two nomograms together.

The final chapter on general theory and analytical developments is most interesting since it serves as an indicator of the state of the development of nomography as an academic discipline. The chapter has two sections, the first is a general study of charts from the point of view of their structure while the second is a study of equations which can be represented by means of a given type of chart. Interesting though the first section is, the subject matter has not proved attractive to subsequent investigators. The objective set by d'Ocagne is to determine and classify all possible methods of representation applicable to equations with n variables. The development concerns itself with the superposition of planes but in practice the number of planes which can be superposed must necessarily be limited. For a limited number of superpositions the theory is helpful but is not of the same calibre as that of the second section.

The objective of the second section is to recognize whether any given equation is associated with a particular type of chart and, if it is, to extract the information necessary to construct that chart. D'Ocagne knows how to solve this problem, in theory at least. His method is to eliminate the arbitrary functions from a general equation. This will result in some partial differential equations which must be satisfied by an equation of that type. The solutions of these equations will lead to the components required to construct the nomogram. This was not the first time that d'Ocagne had expressed this idea, for it occurs in his *Nomographie* of 1891 [**90**]. It is also the basis of the work of Saint-Robert, Massau and Lecornu. The inherent difficulties of this approach are fully recognized, for he points out that the calculations are generally inextricable leading to solutions only in a few special cases.

Thirteen general types of equations are listed as being those which have occurred most frequently in the course of the *Traité*. They are as follows:

I. $f_1(\alpha_1) + f_2(\alpha_2) = f_3(\alpha_3)$

II. $f_1(\alpha_1)f_3(\alpha_3) + f_2(\alpha_2)\phi_3(\alpha_3) + \psi_3(\alpha_3) = 0$

III.
$$\begin{vmatrix} f_1(\alpha_1) & \phi_1(\alpha_1) & \psi_1(\alpha_1) \\ f_2(\alpha_2) & \phi_2(\alpha_2) & \psi_2(\alpha_2) \\ f_3(\alpha_3) & \phi_3(\alpha_3) & \psi_3(\alpha_3) \end{vmatrix} = 0$$

IV. $\lambda_1(\alpha_1)\lambda_2(\alpha_2)f_3(\alpha_3) + \mu_1(\alpha_1)\mu_2(\alpha_2)\phi_3(\alpha_3) + \gamma_1(\alpha_1)\gamma_2(\alpha_2)\psi_3(\alpha_3) = 0$ where λ_i, μ_i, and γ_i are linear functions of α_1 or α_2

V. $f_1(\alpha_1) + f_2(\alpha_2) = f_3(\alpha_3) + f_4(\alpha_4)$

VI. $f_1(\alpha_1)f_2(\alpha_2) + \phi_2(\alpha_2) = f_3(\alpha_3)f_4(\alpha_4) + \phi_4(\alpha_4)$

VII. $(\phi_1(\alpha_1) - \phi_2(\alpha_2))(\phi_3(\alpha_3) - \phi_4(\alpha_4)) + (f_1(\alpha_1) - f_2(\alpha_2))(f_3(\alpha_3) - f_4(\alpha_4)) = 0$

VIII. $f_1(\alpha_1, \beta_1) = f_2(\alpha_2, \beta_2)$

IX. $f_1(\alpha_1, \beta_1) + f_2(\alpha_2, \beta_2) = f_3(\alpha_3, \beta_3)$

X. $f_1(\alpha_1, \beta_1)f_3(\alpha_3, \beta_3) + f_2(\alpha_2)\phi_3(\alpha_3, \beta_3) + \psi_3(\alpha_3, \beta_3) = 0$

XI.
$$\begin{vmatrix} f_1(\alpha_1, \beta_1) & \phi_1(\alpha_1, \beta_1) & \psi_1(\alpha_1, \beta_1) \\ f_2(\alpha_2, \beta_2) & \phi_2(\alpha_2, \beta_2) & \psi_2(\alpha_2, \beta_2) \\ f_3(\alpha_3, \beta_3) & \phi_3(\alpha_3, \beta_3) & \psi_3(\alpha_3, \beta_3) \end{vmatrix} = 0$$

XII. $f_1(\alpha_1) + f_2(\alpha_2) + \ldots + f_n(\alpha_n) = 0$

XIII. $f_1(\alpha_1, \beta_1) + f_2(\alpha_2, \beta_2) + \ldots + f_n(\alpha_n, \beta_n) = 0$

He admits that the method which he has outlined has only been applied to types I and II, but he consoles himself with the remark that they are the most common types in practice. Type III he states can be reduced to certain functional equations which must be identically satisfied. He is, of course, referring to the work of Duporcq which is later acknowledged and reproduced.

The most important section of d'Ocagne is a last chapter that has already been referred to in passing; it concerns itself with the algebraic theory of equations representable by alignment nomograms having three linear supports; i.e.,

those which would later be described as of genus 0 or order 3. The precise problem that he sets himself is this. Suppose that in a general form of an equation,
such as one of those listed above, all of the component functions are algebraic
so that one has a general type of algebraic equation represented by a particular
nomogram; then how, for any given equation, can one form the corresponding components and under what conditions are they real? The reality of the
components is necessary for it to be possible to construct the nomogram. Furthermore, he proposes to seek a solution which will offer the greatest simplicity;
this simplicity he interprets firstly in terms of a nomogram having linear scales
and then considers the transformation of these linear scales into regular scales.
One of d'Ocagne's reasons for this investigation is to reveal the mathematical
depth of a superficially simple problem.

A linear scale is one in which the Cartesian coordinates of a point are given
by

$$x = \frac{m_1\alpha + n_1}{a\alpha + b} \quad \text{and} \quad y = \frac{m_2\alpha + n_2}{a\alpha + b}$$

where α is a parameter and m_1, m_2, n_1, n_2, a, and b are constants. If homogeneous coordinates are used we may write them as

$$x = m_1\alpha + n_1, \quad y = m_2\alpha + n_2, \quad \text{and} \quad t = a\alpha + b.$$

Such nomograms will be either of the type in which the supports are not
concurrent or of the type in which the supports are concurrent. By a homographic transformation the first of these types can be represented by two supports which coincide with the x and y coordinate axes, with the third support
as the line at infinity. After a homographic transformation the second type can
be represented by supports two of which coincide with the x and y coordinate
axes while the third is the bisector of the angle between these axes.

After such a transformation the scales in the case of the non-concurrent
supports may be represented by

$$(i) \ x = m_1\alpha_1 + n_1, \qquad y = 0, \qquad t = p_1\alpha_1 + q_1$$
$$(ii) \ x = 0, \qquad y = p_2\alpha_2 + q_2, \qquad t = m_2\alpha_2 + n_2$$
$$(iii) \ x = p_3\alpha_3 + q_3, \qquad y = m_3\alpha_3 + n_3, \qquad t = 0.$$

The relationship which these scales represent is given by

$$\begin{vmatrix} m_1\alpha_1 + n_1 & 0 & p_1\alpha_1 + q_1 \\ 0 & p_2\alpha_2 + q_2 & m_2\alpha_2 + n_2 \\ p_3\alpha_3 + q_3 & m_3\alpha_3 + n_3 & 0 \end{vmatrix} = 0$$

which can be written as

$$(m_1\alpha_1 + n_1)(m_2\alpha_2 + n_2)(m_3\alpha_3 + n_3) + \qquad (3.1)$$
$$(p_1\alpha_1 + q_1)(p_2\alpha_2 + q_2)(p_3\alpha_3 + q_3) = 0$$

or as

$$M(\alpha_1 + s_1)(\alpha_2 + s_2)(\alpha_3 + s_3) + \qquad (3.2)$$
$$P(\alpha_1 + t_1)(\alpha_2 + t_2)(\alpha_3 + t_3) = 0.$$

Similarly, in the case of the concurrent supports the scales may be represented by

$$(i) \ x = m_1\alpha_1 + n_1, \qquad y = 0, \qquad t = p_1\alpha_1 + q_1$$
$$(ii) \ x = 0, \qquad y = m_2\alpha_2 + n_2, \qquad t = p_2\alpha_2 + q_2$$
$$(iii) \ x = m_3\alpha_3 + n_3, \qquad y = m_3\alpha_3 + n_3, \qquad t = -(p_3\alpha_3 + q_3).$$

which can be written as

$$\frac{p_1\alpha_1 + q_1}{m_1\alpha_1 + n_1} + \frac{p_2\alpha_2 + q_2}{m_2\alpha_2 + n_2} + \frac{p_3\alpha_3 + q_3}{m_3\alpha_3 + n_3} = 0 \qquad (3.3)$$

or

$$\frac{t_1}{\alpha_1 + s_1} + \frac{t_2}{\alpha_2 + s_2} + \frac{t_3}{\alpha_3 + s_3} = N. \qquad (3.4)$$

In the foregoing, α_1, α_2, and α_3 are the variables represented on the scales.

Both (3.1) and (3.3), and their alternative forms, are expressions of the general form,

$$A\alpha_1\alpha_2\alpha_3 + B_1\alpha_2\alpha_3 + B_2\alpha_1\alpha_3 + B_3\alpha_1\alpha_2 + \qquad (3.5)$$
$$C_1\alpha_1 + C_2\alpha_2 + C_3\alpha_3 + D = 0.$$

The problem may now be rephrased.

Given an equation of the form (3.5), under what conditions can it be expressed in one of the forms (3.1) or (3.3) such that the coefficients are real?

The following notion is used:

$$E_i = AC_i - B_j B_k, \qquad F_i = F_0 - 2B_i C_i, \qquad \text{and} \qquad G_i = B_i D - C_j C_k$$

for $F_0 = \sum_{i=1}^{3} B_i C_i - AD$ and $i, j, k = 1, 2, 3$.

D'Ocagne also makes use of a quantity Δ which he describes as the discriminant of (3.5) rendered homogeneous. However, the quantity arises naturally as the discriminant of a quadratic equation, and it will aid understanding if its introduction is deferred.

The basis of d'Ocagne's reasoning is a comparison of (3.5) with an appropriate form of (3.1) or (3.3), depending on the case being investigated. For example, consider the case of non-concurrent supports. Put $\alpha_1 = -s_1$, in (3.2). This gives

$$P(t_1 - s_1)(\alpha_2 + t_2)(\alpha_3 + t_3) = 0. \tag{3.6}$$

The same substitution in (3.5) gives

$$(B_1 - As_1)\alpha_2\alpha_3 + (C_3 - B_2 s_1)\alpha_3 + \tag{3.7}$$
$$(C_2 - s_1 B_3)\alpha_2 + (D - C_1 s_1) = 0.$$

Now if (3.2) and (3.5) are equivalent forms then so also are (3.6) and (3.7). In other words, (3.7) must factorize. The condition for this is that

$$(B_1 - As_1)(D - C_1 s_1) = (C_3 - B_2 s_1)(C_2 - B_3 s_1).$$

Using the notation given above this becomes

$$E_1 s_1^2 + F_1 s_1 + G_1 = 0$$

which is abbreviated $\phi_1(s_1) = 0$. In general, by similar reasoning, $\phi_i(s_i) = 0$ and $\phi_i(t_i) = 0$ for $i = 1, 2, 3$.

Denote the two roots of $\phi_i = 0$ by ρ_i' and ρ_i''. Then (3.2) can be written as

$$M(\alpha_1 + \rho_1')(\alpha_2 + \rho_2')(\alpha_3 + \rho_3') + \tag{3.8}$$
$$P(\alpha_1 + \rho_1'')(\alpha_2 + \rho_2'')(\alpha_3 + \rho_3'') = 0.$$

The form (3.8) can only be obtained if ρ_i' and ρ_i'' are not equal, for if they are for just one value of i then $(\alpha_i + \rho_i')$ is a factor and the required form is lost. It has already been noted that they must be real. Therefore it is necessary that the discriminant Δ of $\phi_i = 0$ is greater than zero; i.e., for

$$\phi_i = E_i s_i^2 + F_i s_i + G_i$$

we have

$$\Delta = F_i^2 - 4E_i G_i > 0.$$

In terms of the coefficients of (3.5), but using F_0, we have

$$\Delta = F_0^2 - 4(B_1 C_1 B_2 C_2 + B_2 C_2 B_3 C_3 +$$
$$B_3 C_3 B_1 C_1 - A C_1 C_2 C_3 - B_1 B_2 B_3 D).$$

Hence for (3.5) to be representable by three linear non-concurrent scales it is necessary that $\Delta > 0$.

It now only remains to find the values of M and P of (3.8), since the values of ρ are given by $\phi_i = 0$ for $i = 1, 2, 3$. These values may be obtained from two of the eight equations of the form

$$MR' + PR'' = K$$

which are obtained by comparing coefficients in (3.5) and (3.8) as follows:

Variable	R'	R''	K
$\alpha_1\alpha_2\alpha_3$	1	1	A
$\alpha_1\alpha_2$	ρ_3'	ρ_3''	B_3
$\ldots$	$\ldots$	$\ldots$	$\ldots$
constant	$\rho_1'\rho_2'\rho_3'$	$\rho_1''\rho_2''\rho_3''$	D

Consider two of these equations:

$$MR_0' + PR_0'' = K_0$$
$$MR_1' + PR_1'' = K_1.$$

These give

$$\frac{P}{R_0' K_1 - R_1' K_0} = \frac{M}{R_1'' K_0 - R_0'' K_1}.$$

Equation (3.8) now becomes

$$(R_1'' K_0 - R_0'' K_1)(\alpha_1 + \rho_1')(\alpha_2 + \rho_2')(\alpha_3 + \rho_3') +$$
$$(R_0'' K_1 - R_1'' K_0)(\alpha_1 + \rho_1'')(\alpha_2 + \rho_2'')(\alpha_3 + \rho_3'') = 0$$

and the non-concurrent problem is solved.

Turning to the problem of concurrent scales, (3.4) can be written in the form,

$$N(\alpha_1 + s_1)(\alpha_2 + s_2)(\alpha_3 + s_3) - t_1(\alpha_2 + s_2)(\alpha_3 + s_3) \qquad (3.9)$$
$$-t_2(\alpha_1 + s_1)(\alpha_3 + s_3) - t_3(\alpha_1 + s_1)(\alpha_2 + s_2) = 0$$

As in the previous case it can be shown that s_1, s_2, and s_3 are respectively roots of the equations $\phi_i = 0$. Once again these roots must be real. However, if the roots of $\phi_i = 0$ are unequal, then the problem reverts to the case of non-concurrent scales; therefore the roots of $\phi_i = 0$ must be equal and $\Delta = 0$ is the condition necessary for (3.5) to be represented by three linear concurrent scales.

If ρ_1, ρ_2, and ρ_3 are roots of $\phi_i = 0$ then (3.9)

$$N(\alpha_1 + \rho_1)(\alpha_2 + \rho_2)(\alpha_3 + \rho_3) - t_1(\alpha_2 + \rho_2)(\alpha_3 + \rho_3) \qquad (3.10)$$
$$-t_2(\alpha_1 + \rho_1)(\alpha_3 + \rho_3) - t_3(\alpha_1 + \rho_1)(\alpha_2 + \rho_2) = 0$$

The coefficients N, t_1, t_2, and t_3 are obtained by comparison of coefficients between (3.10) and (3.5). As before, eight equations are possible. D'Ocagne next investigates how the linear scales may be transformed into regular scales. A general homographic transformation applied to the homogeneous coordinates

$$x = f_i(\alpha_i),$$
$$y = \phi_i(\alpha_i),$$
$$\text{and} \quad t = \psi_i(\alpha_i)$$

is obtained by taking,

$$x = \lambda_1 f_i + \mu_1 \phi_i + \nu_1 \psi_i,$$
$$y = \lambda_2 f_i + \mu_2 \phi_i + \nu_2 \psi_i,$$
$$\text{and} \quad t = \lambda_3 f_i + \mu_3 \phi_i + \nu_3 \psi_i$$

in which

$$\begin{vmatrix} \lambda_1 & \mu_1 & \nu_1 \\ \lambda_2 & \mu_2 & \nu_2 \\ \lambda_3 & \mu_3 & \nu_3 \end{vmatrix} \neq 0$$

A regular scale may be obtained by making the appropriate t constant and different from zero. In other words, in $\lambda_3 f_i + \mu_3 \phi_i + \nu_3 \psi_i$ the term in α_i must be zero and the constant term different from zero. Applied to the case of the non-concurrent scales the general homographic transformation yields,

$$(i) \; x = (\lambda_1 m_1 + \gamma_1 p_1)\alpha_1 + \lambda_1 n_1 + \gamma_1 q_1,$$
$$y = (\lambda_2 m_1 + \gamma_2 p_1)\alpha_1 + \lambda_2 n_1 + \gamma_2 q_1,$$
$$t = (\lambda_3 m_1 + \gamma_3 p_1)\alpha_1 + \lambda_3 n_1 + \gamma_3 q_1;$$
$$(ii) \; x = (\mu_1 p_2 + \gamma_1 m_2)\alpha_2 + \mu_1 q_2 + \gamma_1 n_2,$$
$$y = (\mu_2 p_2 + \gamma_2 m_2)\alpha_2 + \mu_2 q_2 + \gamma_2 n_2,$$
$$t = (\mu_3 p_2 + \gamma_3 m_2)\alpha_2 + \mu_3 q_2 + \gamma_3 n_2;$$
$$(iii) \; x = (\lambda_1 p_3 + \mu_1 m_3)\alpha_3 + \lambda_1 q_3 + \mu_1 n_3,$$
$$y = (\lambda_2 p_3 + \mu_2 m_3)\alpha_3 + \lambda_2 q_3 + \mu_2 n_3,$$
$$t = (\lambda_3 p_3 + \mu_3 m_3)\alpha_3 + \lambda_3 q_3 + \mu_3 n_3.$$

The conditions for regular scales are therefore,

$$\lambda_3 m_1 + \gamma_3 p_1 = 0 \quad \text{with} \quad \lambda_3 n_1 + \gamma_3 q_1 \neq 0,$$
$$\mu_3 p_2 + \gamma_3 m_2 = 0 \quad \text{with} \quad \mu_3 q_2 + \gamma_3 n_2 \neq 0, \tag{3.11}$$
$$\text{and} \quad \lambda_3 p_3 + \mu_3 m_3 = 0 \quad \text{with} \quad \lambda_3 q_3 + \mu_3 n_3 \neq 0.$$

The problem then is to satisfy as many of the conditions (3.11) as possible. Each one that is satisfied gives a regular scale. If they are all to be satisfied then it will not be possible for more than one of λ_3, μ_3, and ν_3 to be zero.

In order to have three regular scales the following equations must be consistent, that is

$$\lambda_3 m_1 + \nu_3 p_1 = 0,$$
$$\mu_3 p_2 + \nu_3 p_2 = 0,$$
$$\text{and} \quad \lambda_3 p_3 + \mu_3 m_3 = 0$$

or in other words

$$\begin{vmatrix} m_1 & 0 & p_1 \\ 0 & p_2 & m_2 \\ p_3 & m_3 & 0 \end{vmatrix} = 0,$$

or, finally,

$$m_1 m_2 m_3 + p_1 p_2 p_3 = 0.$$

Comparing (3.1) with (3.5) gives $A = 0$. This condition is necessary but not sufficient.

Since conditions (3.11) only contain elements from the third row of the determinant

$$H = \begin{vmatrix} \lambda_1 & \mu_1 & \gamma_1 \\ \lambda_2 & \mu_2 & \gamma_2 \\ \lambda_3 & \mu_3 & \gamma_3 \end{vmatrix}$$

it follows that the first two rows may be chosen arbitrarily provided that $H \neq 0$.

If, for example, $\lambda_3 \neq 0$ then we can take $\lambda_1 = 0$, $\mu_1 = 0$, $\gamma_1 = 1$, $\lambda_2 = 0$, $\mu_2 = 1$, and $\gamma_2 = 0$, in which case $m_1 m_2 m_3 = -p_1 p_2 p_3$ so that

(i) $x = p_1 \alpha_1 + q_1,$ $y = 0,$ $t = \lambda_3 n_1 + \gamma_3 q_1;$

(ii) $x = m_2 \alpha_2 + n_2,$ $y = p_2 \alpha_2 + q_2,$ $t = \mu_3 q_2 + \gamma_3 n_2;$

(iii) $x = 0,$ $y = m_3 \alpha_3 + n_3,$ $t = \lambda_3 n_1 + \mu_3 n_3.$

Here the scale for α_1 is the x axis, for α_3 the y axis and for α_2 the straight line $\gamma_3 x + \mu_3 y = 1$. In the case of concurrent scales the same reasoning applies, giving as conditions for regular scales the following:

$$\lambda_3 m_1 + \gamma_3 p_1 = 0 \quad \text{with} \quad \lambda_3 n_1 + \gamma_3 q_1 \quad \neq 0,$$
$$\mu_3 m_2 + \gamma_3 p_2 = 0 \quad \text{with} \quad \mu_3 n_2 + \gamma_3 q_2 \quad \neq 0,$$
$$\text{and} \quad (\lambda_3 + \mu_3) m_3 - \gamma_3 p_3 = 0 \quad \text{with} \quad (\lambda_3 + \mu_3) n_3 - \gamma_3 q_3 \quad \neq 0.$$

If all three scales are regular then the relationship

$$m_1 m_2 p_3 + m_1 p_2 m_3 + p_1 m_2 m_3 = 0$$

must hold. If it does, comparing (3.3) and (3.5) gives $A = 0$. As in the first case this condition is necessary but not sufficient; arbitrary values may be given to the first two rows of H, provided $H \neq 0$.

An analysis of the various possible cases of both types of system leads to the Table 3.1, in which $+$ denotes a positive quantity, $\emptyset$ a non-zero quantity and 0 a zero quantity.

Δ	E_i	E_i	E_i	A	Regular Scales
$+$	$\emptyset$	$\emptyset$	$\emptyset$	$\emptyset$	α_i, α_j
$+$	$\emptyset$	$\emptyset$	$\emptyset$	0	$\alpha_i, \alpha_j, \alpha_k$
$+$	$\emptyset$	$\emptyset$	0	$\emptyset$	α_i, α_j
$+$	$\emptyset$	0	0	$\emptyset$	α_i, α_k
$+$	$\emptyset$	0	0	0	$\alpha_i, \alpha_j, \alpha_k$
$+$	0	0	0	$\emptyset$	α_j
$+$	0	0	0	0	$\alpha_i, \alpha_j, B_i \neq 0$
0	$\emptyset$	$\emptyset$	$\emptyset$	$\emptyset$	α_i, α_j
0	$\emptyset$	$\emptyset$	$\emptyset$	0	$\alpha_i, \alpha_j, \alpha_k$
0	$\emptyset$	$\emptyset$	0	$\emptyset$	α_i, α_j
0	$\emptyset$	0	0	$\emptyset$	α_j, α_k
0	0	0	0	0	$\alpha_i, \alpha_j, \alpha_k$

TABLE 3.1.

By way of illustration consider the following cases. If

$$\alpha_1 \alpha_2 - \alpha_3 = 0,$$

then we have

$$F_0 = -1, F_1 = -1, E_1 = 0, C_1 = D, A = 0, \Delta = 1, E_2 = 0, E_3 = 0.$$

It can therefore be represented by three linear, non-concurrent scales of which two are regular.

$$\frac{1}{\alpha_1} + \frac{1}{\alpha_2} = \frac{1}{\alpha_3}.$$

Writing this as $\alpha_2\alpha_3 + \alpha_1\alpha_3 - \alpha_1\alpha_2 = 0$ we have

$$F_0 = 0, \qquad F_1 = 0, \qquad G_1 = 0, \qquad \Delta = 0$$
$$E_1 = 1, \qquad E_2 = 1, \qquad E_3 = 1, \qquad A = 0.$$

It can be represented by three linear regular concurrent scales.

It is of interest to note that the case in which $\Delta < 0$ was examined very shortly after the publication of d'Ocagne's *Traité*. In November 1900, G. Fontené gave a non-algebraic transformation which changed the value of the discriminant [45]. Starting with (3.5) he made algebraic substitutions of the form $\alpha_1 = \alpha_1' \neq a$ and obtained a reduced form for variables X, Y, and Z, which are linked α_1, α_2, and α_3 by combinations of the coefficients of (3.5). The reduced form is,

$$\varepsilon XYZ + \varepsilon(YZ + ZX + XY) + (X + Y + Z) + 1 = 0$$

where ε is $+1$ or -1 according to the value of the product $B_1B_2B_3D$. If $\varepsilon = -1$, the reduced form is,

$$XYZ + (YZ + ZX + XY) - (X + Y + Z) - 1 = 0$$

with discriminant $\Delta = -16$.

However, the substitutions $X = \tanh U$, $Y = \tanh V$, and $Z = \tanh W$ where $U + V + W = 0$ leads to the equality $X + Y + Z = -XYZ$. This changes the reduced form to,

$$2XYZ + (YZ + ZA + XY) - 1 = 0$$

which has a discriminant $\Delta = 0$.

The final section of the *Traité* deals with the representation of equations of the form

$$A_1\alpha_1^2 + A_2\alpha_2^2 + A_3\alpha_3^2 + 2B_1\alpha_2\alpha_3 + 2B_2\alpha_3\alpha_1 +$$
$$2B_3\alpha_1\alpha_2 + 2C_1\alpha_1 + 2C_2\alpha_2 + 2C_3\alpha_3 + D = 0$$

by means of straight lines and intersecting circles. It is interesting but not as important as the section dealing with regular scales and certainly not as important as the problem to which d'Ocagne might have addressed himself, namely the representation of a relation between three variables by his form III;

i.e.,

$$\begin{vmatrix} f_1(\alpha_1) & \phi_1(\alpha_1) & \psi_1(\alpha_1) \\ f_2(\alpha_2) & \phi_2(\alpha_2) & \psi_2(\alpha_2) \\ f_3(\alpha_3) & \phi_3(\alpha_3) & \psi_3(\alpha_3) \end{vmatrix} = 0.$$

It must be a matter for regret that d'Ocagne did not choose to develop the work of Saint-Robert, Massau, Lecornu and Duporcq. He might have anticipated Grönwall or Kellogg, whose work is discussed later.

Finally, before leaving the *Traité*, it is worth commenting on a footnote which appears on page 209, for it may throw light on how some of d'Ocagne's ideas crossed the Channel. He mentions that he had visited Professor C.V. Boys at the latter's laboratories in the College of Science, London, during May 1896. During this visit Boys stated how useful he had found logarithmic graph paper to be. D'Ocagne clearly means the Royal College of Science, now a part of Imperial College, and we can be fairly certain that Professor Boys received in return advice on the merits of nomography.

3.2. Hilbert's Thirteenth Problem

At the Paris International Congress of 1900 David Hilbert called the attention of mathematicians to twenty three outstanding mathematical problems ([**56**], [**57**]). Problem number thirteen was described as the "Impossibility of the solution of the general equation of the seventh degree by means of functions of only two arguments", and was expressed in nomographic terms.

To understand the problem in the form in which it was presented it is necessary to start with d'Ocagne's concept of points with two dimensions which he introduced in his *Nomographie* [**90**], as doubly isoplethé points and developed in his *Traité* [**92**]. The idea is quite simple and has been met with before. Suppose that the functions of α_1 and β_1, $x = f_1(\alpha_1, \beta_1)$ and $y = \phi_1(\alpha_1, \beta_1)$ have α_1 and β_1 successively eliminated between them. The result is two systems of curves, one having α_1 as parameter and the other β_1, which are given by $F(x, y, \alpha_1) = 0$ and $G(x, y, \beta_1) = 0$ respectively. By taking sets of values for α_1 and β_1 network similar to that of Figure 3.1 can be constructed. In the plane of the network every point has associated with it a pair of values (α_1, β_1) which satisfy simultaneously $x = f_1((\alpha_1, \beta_1)$ and $y = \phi_1((\alpha_1, \beta_1)$.

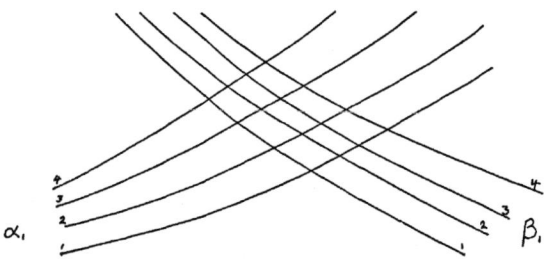

FIGURE 3.1

This idea can be extended so that if, for example, there are three such networks for which

$$(i) \ x = f_1(\alpha_1, \beta_1), \qquad y = \phi_1(\alpha_1, \beta_1),$$
$$(ii) \ x = f_2(\alpha_2, \beta_2), \qquad y = \phi_2(\alpha_2, \beta_2),$$
$$\text{and} \quad (iii) \ x = f_3(\alpha_3, \beta_3), \qquad y = \phi_3(\alpha_3, \beta_3)$$

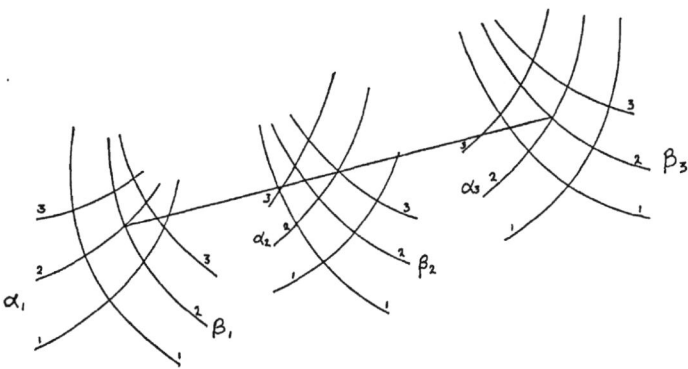

$$\text{FIGURE 3.2}$$

and if they can be used as an alignment nomogram as shown in Figure 3.2 then the equation of which the nomogram is a solution is given by

$$\begin{vmatrix} f_1(\alpha_1,\beta_1) & \phi_1(\alpha_1,\beta_1) & 1 \\ f_2(\alpha_2,\beta_2) & \phi_2(\alpha_2,\beta_2) & 1 \\ f_3(\alpha_3,\beta_3) & \phi_3(\alpha_3,\beta_3) & 1 \end{vmatrix} = 0.$$

There is, of course, no reason why such a system should be limited to only three sets of intersection lines, or indeed, that a particular set of values, say α_1, should not be identical with those of, say, α_2.

Presumably with these ideas in mind, Hilbert points out that a large class of functions of three or more variables can be represented by the above principle alone for, as the translation of this problem by the American Mathematical Society states,

> "... namely all those which can be generated by forming first a function of two arguments, then equating each of these arguments to a function of two arguments, next replacing each of these arguments in their turn by a function of two arguments, and so on, regarding as admissible any finite number of insertions of functions of two arguments."

As an example of a member of this class, Hilbert cites every rational function of any number of arguments as it can be generated by the processes of addition, subtraction, multiplication and division, each of these processes producing a function of only two arguments, It is now a small step to consider the roots of those equations which may be solved by radicals "in the natural realm of rationality" to use the words of the American translation. Such roots demand the four operations of arithmetic together with the extraction of roots, this last process being a function of one argument only, and we see that the roots of such equations also belong to the class of functions under consideration.

Turning his attention to particular equations he considers the general equations of the fifth and sixth degrees. There is a process known as the Tschirnhaus transformation, involving the extraction of roots, which may be used to transform equations. The following theorem applies.

By means of a Tschirnhaus transformation whose coefficients involve a cube root and three square roots, any equation of degree n in x can be transformed into an equation of degree n in y in which the coefficients of y^{n-1}, y^{n-2}, and y^{n-3} are are zero. A good account of the Tschirnhaus transformation may be found in [26]. The general equations of the fifth and sixth degrees may therefore be expressed as,

$$y^5 + py^2 + qy + 1 = 0$$

in which the coefficients of y^{n-1} and y^{n-2} only have been reduced to zero, and

$$y^6 + py^2 + qy + 1 = 0$$

in which the coefficients of y^{n-1}, y^{n-2}, and y^{n-3} have been reduced to zero.

Both of these equations now have coefficients which depend upon only two parameters, p and q, and can be solved by nomography. Indeed, in 1884 d'Ocagne had produced the first alignment nomogram the subject of which was the cubic equation $x^3 + px + q = 0$ which, in principle, is no different from the equations given above Figure 2.16. Hilbert now considers the equation of the seventh degree. Making use of the Tschirnhaus transformation he need only consider the form

$$x^7 + px^3 + qx^2 + rx + 1 = 0$$

about which he makes the following conjecture:

"It is probable that the root of the equation of the seventh degree is a function of its coefficients which does not belong to this class of functions capable of nomographic construction; i.e., that it cannot be constructed by a finite number of insertions of functions of two arguments. In order to prove this, the proof would be necessary that the equation of the seventh degree $x^7 + px^3 + qx^2 + rx + 1 = 0$ is not solvable with the help of any continuous functions of only two arguments."

Hilbert concludes the statement of his problem by saying that he has satisfied himself by a rigorous process that there exist analytical functions of three arguments p, q, and r, which cannot be obtained by a finite chain of functions of only two arguments.

It is not possible to say when d'Ocagne learned of Hilbert's problem, he may have been present at the congress, but his response was swift [**93**]. On September 17, 1900 he presented a paper to the Paris Academy of Sciences on the nomographic resolution of the equation of the seventh degree. He begins with a reference to Hilbert's problem and continues with a very brief comment on his idea of "points with two dimensions" recognizing that a chain of such networks is the solution sought by Hilbert. However he does not pursue this line but introduces the idea of a moveable element, not a new idea since it is a necessary part of Lallemand's hexagonal nomogram. The point being made by d'Ocagne is that in the method of aligned points we have a moveable straight line as a necessary component and that its use is equivalent to the introduction of a system of lines in two dimensions. However this is not so in his solution of the seventh degree equation since it is obtained by the moveable straight line intersecting three scales and in this case the line is a necessary component of a function of the three variables p, q, and r.

Briefly, d'Ocagne's method is as follows. He starts with an equation of the form
$$x_1 + px_2 + qx_3 + rx_4 = 0$$
in which the x_i's are functions of x and p, q, and r may take any values within given ranges.

This equation can be expressed in the determinant form

$$\begin{vmatrix} -1 & q & 1 \\ 1 & r & 1 \\ \dfrac{x_4 - x_3}{x_4 + x_3} & -\dfrac{x_1 + px_2}{x_4 + x_3} & 1 \end{vmatrix} = 0$$

and hence as an alignment nomogram of three scales

(i) $\xi = -1$, $\qquad\qquad\qquad\qquad$ $\eta = q$;

(ii) $\xi = 1$, $\qquad\qquad\qquad\qquad$ $\eta = r$;

(iii) $\xi = \dfrac{x_4 - x_3}{x_4 + x_3}$, $\qquad\qquad\quad$ $\eta = -\dfrac{x_1 + px_2}{x_4 + x_3}$.

In the case of $x^7 + px^3 + qx^2 + rx + 1 = 0$ the following substitutions are made

$$x_1 = x^7 + 1, \quad x_2 = x^3, \quad x_3 = x^2, \quad \text{and} \quad x_4 = x$$

in which case scale (iii) becomes,

$$\xi = \frac{1 - x}{1 + x} \quad \text{and} \quad \eta = -\frac{x^7 + px^3 + 1}{x + x^2}$$

and if x is eliminated between them,

$$\eta = -\frac{\left(\frac{1-\xi}{1+\xi}\right)^7 + \left(\frac{1-\xi}{1+\xi}\right)^3 + 1}{\left(\frac{1-\xi}{1+\xi}\right) + \left(\frac{1-\xi}{1+\xi}\right)^2}.$$

Thus, for each value of p there is a curve in the (ξ, η) plane. Additionally, the ξ axis is a scale of x since

$$\xi = \frac{1 - x}{1 + x}.$$

From scales (i) and (ii) it is seen that the values of q and r lie on lines parallel to the η axis and at a unit distance from it on either side.

Figure 3.3 illustrates the scheme. In the discussion scale factors have been ignored although they would obviously be very important for a sensible and accurate nomogram. Ingenious though d'Ocagne's work is, it is no answer to the problem posed by Hilbert. Although Hilbert posed his problem in nomographic terms it is related to the much wider problem of the complexity of functions. There does not appear to have been much progress made towards resolving it

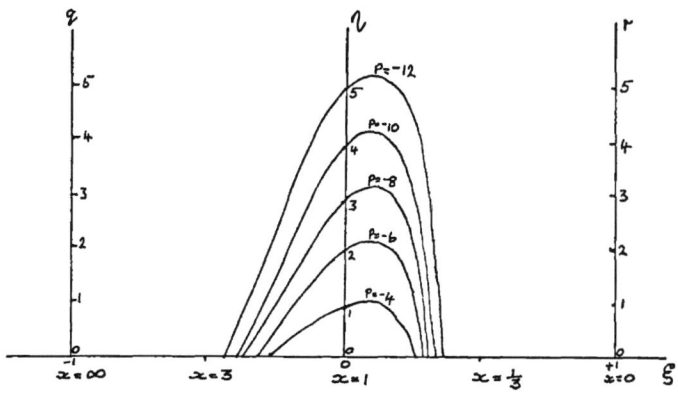

FIGURE 3.3

until the mid-1950's. In 1956 A.N. Kolmogorov proved that every continuous function of n variables can be represented in the form of a superposition of continuous functions of three variables [62]. The following year, V.I. Arnol'd was able to modify this theorem reducing the number of variables from three to two [4]. Also in 1957, Kolmogorov published a theorem, one consequence of which is that, within appropriate limits, the function $f(x_1, x_2, x_3)$ may be expressed as,

$$f(x_1, x_2, x_3) = \sum_{q=1}^{7} h_q(\phi_q(x_1, x_2), x_3)$$

where all the functions are continuous [63].

These interesting results do not dispose of Hilbert's problem. The distinction must be made between algebraic functions and continuous functions of two variables. Hilbert's problem is algebraic in origin, since it arises out of attempts to eliminate as many coefficients as possible from polynomial equations

$$\sum_{k=0}^{n} a_k x^k = 0,$$

but he recognizes its more general application and appears to expect that the seventh degree equation could not be solved even by continuous functions of two variables. The work of Kolmogorov and Arnol'd show that this last supposition was wrong but Hilbert's original problem remains unsolved. It is still not known whether the equation

$$x^7 + px^3 + qx^2 + rx + 1 = 0$$

can be solved by a finite number of superpositions of algebraic functions of two variables.

3.3. The Spread of Nomographic Ideas

The story of the development of nomography so far has been the story of the discipline in the hands of specialists. To be sure, most of these specialists were engineers who developed their ideas in order to achieve practical ends but the important fact is that they were developers; no case of an engineer or scientist taking these ideas and suggesting how they might be applied to his own discipline has yet been cited. The earliest such cases of which I am aware date from around the turn of the century, some before but most after, but there may be other cases which I have not found. There is evidence in many of these cases that it was the works of d'Ocagne, usually his *Traité*, which were responsible for the spread of these ideas. It is of interest to assemble the available evidence to see when, and for what applications, the ideas of nomography were transmitted to those who would wish primarily to be users.

In Britain, the first indication of the use of geometric methods in computation in the new century is rather disappointing. In 1903 the Minutes of the Proceedings of the Institution of Civil Engineers contained a short item by R.S. Scholfield on "The use of logarithmic scales in plotting curves" [124]. In this article the author states the desirability of replacing curves by straight lines whenever this is possible and makes the following suggestions, apparently unaware that some of them had appeared in print as much as fifty years earlier. For expressions of the form $ab = c$ he suggests plotting $\log a + \log b = \log c$ and for $xR = kv^2$ the suggestion is to plot the v's to a scale of squares to obtain radial straight lines passing through the origin. His third suggestion, in a case where curves approximate to hyperbolæ of the form $xy =$ constant, is to use polar coordinates with a base line divided logarithmically. Sound though these ideas are it is regrettable that they needed to be published for the Institution of Civil Engineers as late as 1903.

Perhaps the first published account in English of nomographic methods was a series of articles by John B. Peddle which appeared in *The American Machinist* during 1908 [109]. The purpose behind these articles was clearly a practical one as the choice of publication suggests. Undoubtedly many American engineers encountered geometric computation for the first time on reading them and it is very possible that the same is true of British engineers for the publication seems to have had a wide circulation.

During the following year there appeared in the Journal of the Royal Artillery a very short paper describing a "Scale for the graphic calculation of deflection and angle of sight problems," by Captain R.K. Hezlet, R.A. [52]. The scale was an alignment nomogram which had as variables the range R, height H, and angle S connected by the formula $H = R \sin S$. The limits for the variables conformed to the requirements of artillery at that time. This particular nomogram was apparently on sale, for Hezlet states the following:

> "The scale is printed on a stout card $6'' \times 4\frac{1}{2}''$ and the edges of the back of the card have been graduated with useful scales for map reading purposes, one edge being left blank so that it can be graduated with a degree scale to suit the user's length of arm.
>
> The cards may be obtained from Messrs. W. Watson & Sons, 313, High Holborn, London."

The following year, 1910, Hezlet wrote a longer paper for the *Journal of the Royal Artillery* entitled "The Graphic Representation of Formulae" [53]. He begins this paper with the statement that it is the outcome of a study of two works of d'Ocagne, the *Traité* and "Calcul graphique et Nomographie," the latter having been published in 1908. Pointing out that there did not appear to be an English equivalent of the French words 'Nomographie' and 'Nomogramme', he coins the word *nomogram* to describe the graphical chart and this appears to be the first time that the word nomogram is used in a British publication. The substance of the paper is the application of nomograms to some formulae in ballistics and in particular of alignment nomograms which Hezlet prefers. He makes no attempt to explain the theory behind such nomograms, referring the curious to d'Ocagne's works, but gives rules for the construction of certain nomographic types, illustrating them with examples from ballistics. The formula types dealt with are

$$f_1(z_1) + f_2(z_2) + f_3(z_3) = 0,$$
$$f_1(z_1) + f_2(z_2)f_3(z_3) = 0,$$
$$\text{and} \quad f_1(z_1)g_3(z_3) + f_2(z_2)h_3(z_3) + f_3(z_3) = 0.$$

Those requiring methods for formulae involving four, five or six variables are again referred to the works of d'Ocagne.

The importance of Hezlet's two papers lies in the fact that they mark the introduction of d'Ocagne's ideas into Britain. Hezlet has the honor of being the pioneer of nomography in this country but he always disclaimed originality, being fulsome in his praise of d'Ocagne. It is not known whether d'Ocagne and Hezlet ever met, they may have done for Hezlet served in France during 1915 but the conditions then would not have been conducive to such a meeting.

In 1911 another soldier, this time from the Corps of Royal Engineers, showed a familiarity and appreciation of d'Ocagne's works. Captain C.E.P. Sankey wrote a paper with the title "Moving loads on military bridges" which included several pages on "The graphical representation of formulae" [**123**]. The following is taken from his paper,

> "The subject of graphical charts in general is a most fascinating one, and many books and articles have been written on it. Among these may be mentioned *Traité de Nomographie* by Maurice d'Ocagne, a work that is most exhaustive in its treatment;..."

Other authors mentioned are Peddle and Scholfield. Thus it is clear that d'Ocagne's work was known at least to officers of the two technical corps of the British army by 1911. In his general remarks, Sankey complains that graphical charts are not sufficiently used on service and gives reasons why the military should be interested in them, making the point that, in general, a graphical chart is only an economy if the time required for its preparation is less than the aggregate time necessary for the separate calculations which it will replace, he observes that on military service priorities are different and that a small saving of time, even on only one occasion, can be very cheaply purchased by the time required to prepare such a chart. The brief review of nomographic methods given by Sankey is concise end comprehensive and is designed to whet the appetite of the reader; it includes the ideas of anamorphosis, of intersection nomograms and of alignment nomograms and is accompanied by many charts prepared by the author. An appendix on alignment nomograms, which gives a brief introduction to parallel coordinates, is included.

If the ideas of nomography were to gain wider acceptance in Britain then it would be necessary for a text on the subject to be published. Such a text was written by Hezlet. It was published by the Royal Artillery Institution in March 1913 and cost two shillings and sixpence [55]. It is a very lucid and concise little book which shows the author to be a competent mathematician although he disclaims all originality except in so far as the examples are concerned. All aspects of nomography likely to be wanted by scientists or engineers are included with enough theoretical support to satisfy the theoretically inclined, but those not so inclined are recommended to disregard theory entirely at first and go straight to the examples. It is a measure of the ability of the author that it is quite possible to use the book in this way.

In spite of this excellent little book, British ignorance of nomography was apparently still widespread in 1920. This was so even in Hezlet's own organization, for he found it necessary to write a short note in the Royal Artillery Journal with the title "What is a Nomogram?" [54]. In this he takes a calculation, which would have been familiar to his readers, concerning wind corrections to the line of fire and draws both an intersection nomogram and an alignment nomogram for it. He makes two points; firstly, that a nomogram is more nearly foolproof than a normal calculation and secondly, that an alignment nomogram is to be preferred to an intersection nomogram.

Perhaps it was this widespread ignorance of nomography that prompted a member of the British academic establishment to publish a book on the subject, for in 1920 there appeared *A First Course in Nomography* by Selig Brodetsky, then Reader but later Professor of Mathematics at the University of Leeds [12]. The book is no more than it claims to be; i.e., a first course, but it must have been very useful to the engineer who wanted an introduction to the techniques.

It was not until 1932 that anything like a text book on nomography was published in Britain and even then it could not compare in any real sense with d'Ocagne's *Traité*. This was *The Nomogram* by Allcock and Jones which, through its various editions, became the standard British text [1].

Before examining the rather more sparse evidence of the spread of nomographic ideas to other countries, it is worth looking at some aspects of Hezlet's life. It is strange that those men who have been concerned with nomography and whose lives are a matter of public record have been unusually versatile men, as we have seen in the cases of Lalanne and d'Ocagne. Hezlet is no exception. A professional soldier, he was an expert in ballistics who spent a great part of his career in research and experimentation on guns and ammunition. During the First World War he was twice mentioned in dispatches and awarded the D.S.O. before his specialized knowledge caused him to be returned to Britain to a post at the Ministry of Munitions. He reached the rank of Major General and amongst his appointments was that of Commandant of the Military College of Science. This would be enough for most men but in retirement Hezlet took the normal undergraduate course at the Royal Veterinary College, Edinburgh, and, at the age of 67, qualified as a veterinary surgeon. He died at the age of 83.

In passing we may note that Brodetsky was no less versatile. Born in Russia and brought up in East London, he won his way to Trinity College, Cambridge, and was Senior Wrangler in the Mathematics Tripos of 1908. His mathematical career was or of considerable success but in addition he was dedicated to the Zionist cause. Amongst the appointments which he held in this connection were: member of the Board of governors of the Hebrew University, member of the executive of the Zionist World Organization and President of the Board of Deputies of British Jews.

In France one would have expected d'Ocagne's ideas to have been accepted speedily, not only because he wrote in French but also because many of those most likely to use his techniques might well have been taught by him. This had happened in the case of a naval officer, one Lieutenant Perret, who learned about alignment nomograms by attending classes given by d'Ocagne at l'École Polytechnique in 1894 − 1895. In 1904, Perret published a paper on the application of alignment nomograms to problems in Nautical Astronomy and, in the following year, he addressed the French Academy for the Advancement of Science at its meeting in Cherbourg on the same subject [112]. His paper is a substantial one having, as Perret states, "...the purpose of interesting our friends in the use of a procedure which can render to them real service." Stating that nautical problems are normally solved by the use of tables and that these are of great convenience when the related equation contains only two or three variables, he points out that an increase in the number of variables can greatly diminish this convenience. As an example he cites the method of determining the azimuth of a star, when the latitude, hour angle and declination are known, for which the tables consist of several volumes. He chooses this problem as his first example. However, before he deals with this he gives a short explanation of the theory of alignment end of parallel coordinates and this is clearly intended

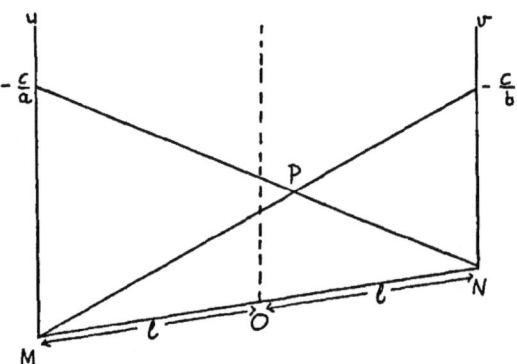

FIGURE 3.4

for the user rather than the theorist. His approach is a very simple one. It is based on comparing a given equation with the equation $au + bv + c = 0$, which represents a point in the case where u and v are parallel coordinates. In Figure 3.4, if M is the point from which u is measured and N that from which v is measured, then P is the point having $au + bv + c = 0$ as its equation.

If Cartesian coordinates are superimposed on the diagram in such a way that the x axis coincides with MN, the origin 0 with the mid point of MN, and the y axis parallel to the u and v axes, then P is given by

$$x = \ell\frac{b - a}{b + a} \quad \text{and} \quad y = -\frac{c}{b + a}$$

where MN is of length 2ℓ.

Thus, for an equation of the form,

$$\psi_1(\alpha_3)f_1(\alpha_1) + \psi_2(\alpha_3)f_2(\alpha_2) + \psi_3(\alpha_3) = 0 \tag{3.12}$$

on comparison with $au + bv + c = 0$, we may write,

$$u = f_1(\alpha_1), \quad v = f_2(\alpha_2) \tag{3.13}$$

and

$$a = \psi_1(\alpha_3), \quad b = \psi_2(\alpha_3), \quad c = \psi_3(\alpha_3). \tag{3.14}$$

The equations (3.13) give directly the graduations of the parallel axes u and v.

The equations (3.14) lead to the Cartesian coordinates of points in the plane, each point depending on the particular value of α_3. These coordinates are

$$x = \ell \frac{\psi_2(\alpha_3) - \psi_1(\alpha_3)}{\psi_2(\alpha_3) + \psi_1(\alpha_3)} \quad \text{and} \quad y = -\frac{\psi_3(\alpha_3)}{\psi_2(\alpha_3) + \psi_1(\alpha_3)}.$$

It can be seen that if α_3 is eliminated from these coordinates the result is an equation of the form $F(x, y) = 0$. To each point of this curve there will correspond a particular value of α_3.

Thus, three scales, one each for α_1, α_2, and α_3, are obtained quite simply and without having the problem of putting (3.12) into a determinant form, an obvious advantage to the user. An alignment nomogram of the form obtained is shown in Figure 3.5.

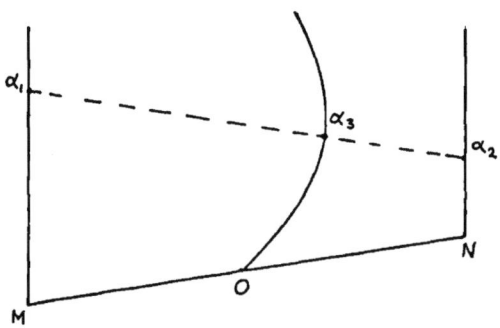

FIGURE 3.5

The azimuth problem has for its equation,

$$\cot A \sin H + \cos L \tan \Theta - \sin L \cos H = 0 \qquad (3.15)$$

in which

        L is the latitude,
        Θ is the angle of declination,
        H is the astronomical hour angle, and
        A is the azimuth of the star.

Comparing (3.15) with $au + bv + c = 0$, we have

$$u = \cot A, \qquad v = \tan \Theta, \qquad (3.16)$$

$$a = \sin H, \qquad b = \cos L, \qquad \text{and} \qquad c = -\sin L \cos H \qquad (3.17)$$

giving

$$x = \ell \frac{\cos L - \sin H}{\cos L + \sin H} \qquad \text{and} \qquad y = \frac{\sin L \cos H}{\cos L + \sin H}. \qquad (3.18)$$

Equations (3.18) give a series of points which depend on two variables, namely L and H, and therefore amount to an intersection nomogram which can, given that L lies between $-\frac{\pi}{2}$ and $\frac{\pi}{2}$, be such that $|x| < \ell$; i.e., it lies between the vertical scales $u = \cot A$ and $v = \tan \Theta$.

The construction of the scales given by (3.18) is not as tedious as it may seem for, by eliminating in turn L and H, we arrive at

$$4\ell y^2 \tan^2 H - x^2 \cos^2 H + 2\ell x(1 + \sin^2 H) - \ell^2 \cos^2 H = 0 \qquad (3.19)$$

$$\text{and} \quad 4\ell y^2 \cot^2 L - x^2 \sin^2 L - 2\ell x(1 + \cos^2 L) - \ell^2 \sin^2 L = 0$$

which represent hyperbolas having the x axis as diameter. The appearance of the nomogram is shown in Figure 3.6 where d represents Θ. The small diagonal scales and horizontal lines cutting them allow use at greater scale ranges without increasing the size of the nomogram.

Perret gives much practical advice on the construction of this nomogram, advice that is to help the navigator construct a nomogram which would be of use at sea.

The second example is concerned with the prediction of occultations. This again is treated in a very full manner, starting from that which would be familiar to the sailor and then splitting the problem into sections which can be dealt with as described in the preceding example.

It is of interest to note that during the same year d'Ocagne was working on the nomographic solution of spherical triangles, a topic of some importance to astronomical navigation and surveying, areas to which nomography was being applied at this time ([95], [96]). Also in France, it is recorded that the physics course at l'École Polytechnique made use of an alignment nomogram to carry out calculations for Van der Waal's gas equation. When the warship Republique was tested, the chief marine engineer carried out many of the necessary calculations by nomograms. In surveying, a work by Capt. de Larminet called *Topographie Pratique* had a supplement which contained eight nomograms dealing with various ordinary calculations of surveying, while Capt. Lelarge, writing in the *Revue de Génie*, gave nomograms relating to telephotographic surveying ([77] and [78]). The world of business was also making use of nomograms; an actuary by the name of Poussin published, in 1904, a series of alignment nomograms dealing with insurance problems [118].

These are but a few of the examples which show how, in the early years of the twentieth century, nomograms, and in particular alignment nomograms,

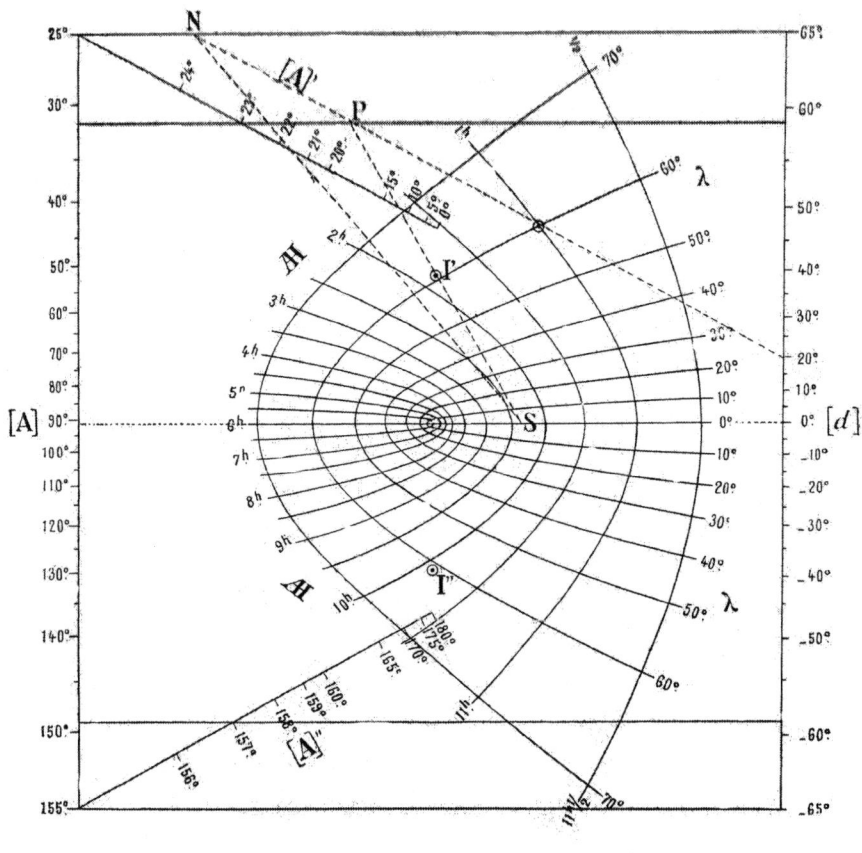

FIGURE 3.6

had become accepted as an important aid to calculation in many branches of knowledge in France.

Those countries in which France had some influence, either through language or by the presence of French citizens, might also be expected to have taken to nomographic ideas. This would certainly have been true of Canada, where one can assume that the French language would not be a great barrier, and of Egypt, where French engineers would have been prominent in the Suez Canal Company.

In Canada, E. Deville, a well known land surveyor who was head of the Geodetic Service of Canada at the time in question, presented a paper to the Royal Society of Canada on the use of nomography to find the altitude and azimuth of the pole star [25]. In many respects this is a parallel paper for surveyors to the one written by Perret for navigators for it is concerned with eliminating calculation from a problem met with continually in the exercise of that particular activity. The paper has a section on the graphic representation of equations which begins with an acknowledgement to d'Ocagne for his "exhaustive investigation of the subject." It is followed by a brief account of that part of the subject required for his purposes; this part is of a higher mathematical standard than the similar part given by Perret. Deville explains how intersection nomograms and alignment nomograms can be connected through the principle of duality and gives the determinant form for an intersecting system of three straight line sets. In view of the fact that in the introduction he makes the remark "some surveyors prefer no calculation whatever," one wonders whether such surveyors got anything from this part of the paper.

Denville's approach to the two particular problems, the azimuth of the pole star and its altitude, is along the same lines as Perret's; i.e., the appropriate formula is put in a form suitable for the precise problem and then compared with $au + bv + c = 0$. Deville differs from Perret by introducing moduli, or scale factors, at this stage. The final nomograms have the appearance of great simplicity.

In Egypt, nomograms were produced by the head of the Irrigation Service for calculations dealing with the flow of water in canals and over weirs [135]. Problems on the strengths of railway bridges were solved by Farid Boulad, an engineer with Egyptian Railways, using alignment nomograms [9]. It must also be noted that the next major advance in nomography was to be contained in a series of articles by Dr. J. Clark written in 1907 when he was Professor of Mathematics at l'École Polytechnique in Cairo. These are examined in the next section.

Nomograms were produced in Italy before the end of the nineteenth century. Professor Gorrieri produced some on problems concerning the strength of loaded beams [50], while Professor Molfino of Livorno carried out a study similar to that of Perret in France [84]. Professor Pesci, also of Livorno, used alignment nomograms to solve problems in naval kinematics [113]. Problems related to artillery were also the subject of Italian nomograms by Ronca [121], and Ricci [119].

In Spain, General Ollero used alignment nomograms to solve problems in ballistics [106].

In Holland, an engineer named Vaes produced alignment nomograms to solve problems arising in the construction of marine boilers [132].

Moving eastwards we find that W. Láska, Professor of Geodesy at the Technical High School, Lviv, in the Ukraine, had used alignment nomograms for topographic calculations by 1905 and was in the process of collaborating with an engineer named F. Ulkowski to produce a collection of technical nomograms [74]. I have not found any trace of this collection.

Nomography was introduced into Russia, according to the Russian Encyclopedia, by M.N. Gersevanov during the period 1906 to 1908, but d'Ocagne records that Colonel Langensheld had used alignment nomograms for ballistic calculations in Russia with the implication that this may have been before 1906. D'Ocagne also acknowledges the work of Gersevanov and credits him with the invention of a new type of nomogram called à points equidistants [98]. I have not found reference to this type of nomogram anywhere else.

It is clear that by the end of the first decade of the twentieth century the discipline of nomography had spread far from its birthplace of France. It is also clear that it was taken up more enthusiastically in some countries than in others. In Britain there seemed to be a reluctance to take up the ideas and they were never to become as important as they deserved to be. The countries which were to contribute most to nomography in the future were Poland end Russia, both of which were very active in the 1950's. The United States also retained an interest in the subject. One can speculate on reasons for this state of affairs but it can only be speculation for there is no real evidence. British engineers appear to have been devoted to their slide rules and this, together with the insular attitude to which Britain has always been prone, was probably enough to confine nomography to a minor role. It is not easy to account for the sustained Russian interest, which extended throughout the 1930's, a seemingly static period elsewhere, except by suggesting that the nomographic approach to problems may have been more in tune with the Russian temperament and outlook. The great interest in Russia and Poland in the 1950's is easier to understand for at this time the West was making great strides forward in the development of electronic computers and the East European countries may well have felt that they were being left behind in the field of computation; to improve nomographic techniques would at least have been an interim measure to narrow the gap. Recent American interest might be just a normal reaction to the Russian interest but this could be an oversimplification; there is universal interest in the problem of complexity of functions, a problem which originated in nomography.

From about 1907 onwards, the publications dealing with nomography at all levels proliferated greatly. A study of the development of nomography can henceforward only take account of those developments which advanced the discipline in some major way or brought out some vital point. It therefore follows that some worthwhile contributions will be left out while some, perhaps less worthwhile, will be included because they illuminate the particular point being made. The task of providing a catalogue of nomograms over the years is a different task from the one in hand.

3.4. The Contribution of Dr. J. Clark

Dr. Clark first presented his ideas on nomography in 1905 to the Cherbourg congress of the French Association for the Advancement of Science. It was also at this congress that Lt. Perret presented his paper on the application of nomography to nautical astronomy. Whereas Perret's work had appeared in print the year before, Clark's did not appear until three years after the conference. When it did appear it was in the Revue de Mécanique and was split into sections, the first appearing in October 1907 and the last in May 1908 ([18], [19], [20], [21]). The Revue de Mécanique is rather an odd choice since the work itself is mathematical; d'Ocagne's introduction begins with the comment that it "diverges somewhat from the ordinary bounds of that received," but it is not known why that publication was chosen. Very little has been recorded about Clark the man. At the time that his articles appeared he was Professor of Mathematics at l'École Polytechnique in Cairo. He does not seem to have been particularly fluent in French since his articles were translated into French by G. Fleuri, another professor in Cairo. D'Ocagne's introduction is curious. While praising Clark's work it nevertheless conveys the impression that most credit is due to d'Ocagne himself. In one sense this is true, for d'Ocagne can rightly be considered the father of the subject, but if Clark's work depends on any one idea, that idea is Soreau's notion of nomographic order. The tendency of d'Ocagne towards self aggrandisement in his later writings is most noticeable. Taken with his apparent plagiarism of the ideas of critical points from Massau and of nomographic order from Soreau, one is left with a feeling of regret that an able man should conduct himself in such a way.

Since the time of Lalanne the idea that a nomogram should contain as many straight line supports as possible had become a guiding principle. In practice, using the classification ideas of Soreau and d'Ocagne, this meant searching for a nomogram of genus $n - 3$ when the equation was of nomographic order n, n being 3, 4, 5, or 6. Clark's important contribution was to take this principle and investigate the construction of nomograms when it was not followed. The result was that he sought nomograms of genus greater than $n - 3$ for equations of nomographic order n.

The quality of this mental process must be compared in nomography with d'Ocagne's work which led to alignment nomograms. In mathematics generally there is a greater precedent for it. Clark's abandonment of what may be called the anamorphosis principle is of the same type of reasoning as that which led to the abandonment of the parallel principle in Geometry, and which in turn led to the discovery of non-Euclidian geometries.

Clark's work begins with an examination of the general equation of the fourth nomographic order which he writes in the form,

$$
\begin{aligned}
&f_3(a_0 f_1 f_2 + a_1 f_1 + a_2 f_2 + a_3) + \\
&\phi_3(b_0 f_1 f_2 + b_1 f_1 + b_2 f_2 + b_3) + \\
&\psi_3(c_0 f_1 f_2 + c_1 f_1 + c_2 f_2 + c_3) = 0.
\end{aligned}
\tag{3.20}
$$

The problem posed is to investigate under what conditions this equation may be represented by a nomogram of the same orders; i.e., by two straight lines and one curve, or in d'Ocagne's classification, of genus 1. It should be noted that if $\psi_3 = 0$ in the general equation then it is reduced to the third order.

As the canonical type of the fourth order equation Clark takes

$$
F_1 F_3 + F_2 \phi_3 = \psi_3.
\tag{3.21}
$$

It is the form given by Cauchy when he considered Lalanne's paper on anamorphosis. The problem is now to find under what conditions it is possible to pass from (3.20) to (3.21), or vice-versa, by a homographic transformation.

If (3.20) is written in the shorter form

$$
f_3 x + \phi_3 y + \psi_3 z = 0,
\tag{3.22}
$$

then Clark's first condition is that a value of λ must exist such that $x + \lambda y$ can be factorized. When the condition that λ should be real is applied, the result goes back to d'Ocagne's work on the third order equation for it is no more than that the discriminant of $f_3 x + \phi_3 y = 0$ should be greater than or equal to zero, It therefore follows that a first condition for the representability of a fourth order equation by a nomogram of the same order is that all equations of the third order obtained by making f_3 or ϕ_3 or ψ_3 equal to zero must be representable by a chart of the same order.

Starting from (3.21), F_1 and F_2 are written as,

$$F_1 = \frac{\chi_1}{\psi_1} \quad \text{and} \quad F_2 = \frac{\chi_2}{\psi_2},$$

these forms being linear functions of f_1 and f_2 respectively. This is necessary since in the canonical form it is these two functions which represent the straight lines. Clark's process for showing when (3.20) and (3.22) are homographic transformations of each other involves at one stage a comparison of equations which indicates that the z of (3.22) must have $\psi_1\psi_2$ as a factor. Furthermore, when this factor is removed an expression linear in ψ_1^{-1} and ψ_2^{-1} is left, unless ψ_1^{-1} and ψ_2^{-1} are constant in which case it is linear in f_1 and f_2. This leads to some practical rules.

- Decompose $x + \lambda y$ into factors. There will at the most be two values of λ.
- If the two values of λ are real and distinct, then

$$x + \lambda_1 y = \Omega_2 \psi_1$$
$$\text{and} \quad x + \lambda_2 y = \Omega_1 \psi_2$$

  and the required factors are either $\psi_1\psi_2$ or $\Omega_1\Omega_2$.
- If division by one of the pairs makes z linear, then the equation is of the required form, if not it is irreducible to that form.

If the two values of λ are real and equal then $x + \lambda y = \psi_1\psi_2$ and the required factors are ψ_1 and ψ_2. If z is rendered linear on division by $\psi_1\psi_2$ then the equation is of the required form, if not it is irreducible. If there are no real values of λ the equation is irreducible.

By way of illustration consider the following equation:

$$Af_3(f_1 + a_1)(f_2 + a_2) +$$
$$B\phi_3(f_1 + b_1)(f_2 + b_2) +$$
$$C\psi_3(f_1 + c_1)(f_2 + c_2) = 0.$$

Firstly note that since x, y, and z are all decomposed into factors then these factors must contain those sought. The equality

$$x + \lambda y = (f_1 + a_1)(f_2 + a_2) + \lambda(f_1 + b_1)(f_2 + b_2)$$

shows at once that the equation is not reducible unless two factors taken from different terms are identical. This however is based on the fact that $x + \lambda y$ cannot be factorized. Suppose that it has factors, say $(f_1 + a_1)$ and $(f_2 + b_2)$, then the coefficient of ψ_3 after division, is

$$\frac{(f_1 + c_1)(f_2 + c_2)}{(f_1 + a_1)(f_2 + b_2)} = \left(1 + \frac{c_1 - a_1}{f_1 + a_1}\right)\left(1 + \frac{c_2 - b_2}{f_2 + b_2}\right)$$

which is only linear in $(f_1 + a_1)^{-1}$ and $(f_2 + b_2)^{-1}$ if $(c_1 - a_1)(c_2 - b_2) = 0$; i.e., if $c_1 = a_1$ or $c_2 = b_2$, which again shows that the equation is irreducible unless two factors taken from different terms are identical.

Of some interest for their generality are two forms given by Clark as examples. The first is

$$f_1 f_3 + f_1 \phi_3 = 1 + k f_1 f_2$$

where k is a constant. He points out that this equation is of the form

$$f_1 c_2 f_3 + f_2 c_1 \phi_3 = 1 + k f_1 f_2$$

where c_1 and c_2 are constants representing degenerate linear functions for f_1 and f_2.

Following the method used above, both sides can be divided by $c_1 c_2$ giving

$$\frac{f_1 f_3}{c_1} + \frac{f_2 \phi_3}{c_2} = \frac{1}{c_1 c_2} + \frac{k f_1 f_2}{c_1 c_2}$$

showing that the right hand side is only linear when $k = 0$. Hence,

$$f_1 f_3 + f_2 \phi_3 = 1 + k f_1 f_2$$

is only reducible to the canonical type (3.21) if $k = 0$.

The second example is

$$f_1 f_3 + f_2 \phi_3 = f_1 f_2 + k.$$

If $k \neq 0$, then putting $m = k^{-1}$ the previous form is obtained. If $k = 0$, then on dividing by $f_1 f_2$,

$$\frac{f_3}{f_2} + \frac{\phi_3}{f_1} = 1$$

is obtained which is the canonical type (3.21). Hence

$$f_1 f_3 + f_2 \phi_3 = f_1 f_2 + k$$

is only reducible to the canonical type (3.21) if $k = 0$.

Clark is able to condense these results into a more manageable form in order to "recognize at first sight whether an equation of the fourth order is representable or not."

He rewrites equation (3.20) in the form

$$f_1 f_2 A_3 + f_1 B_3 + f_2 C_3 + D_3 = 0 \tag{3.23}$$

and concentrates on the coefficients A_3, B_3, C_3, and D_3, where

$$A_3 = a_0 f_3 + b_0 \phi_3 + c_0 \psi_3,$$
$$B_3 = a_1 f_3 + b_1 \phi_3 + c_1 \psi_3, \tag{3.24}$$
$$C_3 = a_2 f_3 + b_2 \phi_3 + c_2 \psi_3,$$
$$\text{and} \quad D_3 = a_3 f_3 + b_2 \phi_3 + c_3 \psi_3.$$

Firstly, if any one of the coefficients is zero, then equation (3.24) reduces to forms which can be seen to be equivalent to the canonical form (3.21).

If all the coefficients are different from zero, then an examination of their forms shows that a linear relationship must exist between them; i.e.,

$$a A_3 + b B_3 + c C_3 + d D_3 = 0.$$

This means that D_3 can be eliminated from (3.23) provided that $d \neq 0$, giving

$$\left(f_1 f_2 - \frac{a}{d} \right) A_3 + \left(f_1 - \frac{b}{d} \right) B_3 + \left(f_2 - \frac{c}{d} \right) C_3 = 0. \tag{3.25}$$

By applying the method described earlier, for representability it is necessary that

$$\frac{f_1 f_2 - \frac{a}{d}}{\left(f_1 - \frac{b}{d} \right) \left(f_2 - \frac{c}{d} \right)}$$

should be linear in $(f_1 - \frac{b}{d})^{-1}$ and $(f_2 - \frac{c}{d})^{-1}$. Put $X_1 = f_1 - \frac{b}{d}$ and $X_2 = f_2 - \frac{c}{d}$, then

$$\frac{(X_1 + \frac{b}{d})(X_2 + \frac{c}{d}) - \frac{a}{d}}{X_1 X_2}$$

must be linear in X_1^{-1} and X_2^{-1}; i.e.,

$$\frac{bc}{d^2} - \frac{a}{d} = 0.$$

or $ad - bc = 0$. If $d = 0$, then some other coefficient is eliminated, say B_3 if $b \neq 0$, and the argument repeated, giving the same result. This result is quite general.

Returning to equation (3.25) and making the substitution $X_1 = f_1 - \frac{b}{d}$ and $X_2 = f_2 - \frac{c}{d}$ we have

$$\left(\left(X_1 + \frac{b}{d} \right) \left(X_2 + \frac{c}{d} \right) - \frac{a}{d} \right) A_3 + X_1 B_3 + X_2 C_3 = 0;$$

i.e.,

$$X_1 f_3 + X_2 \phi_3 = X_1 X_2 + \frac{bc - ad}{d^2}$$

which reinforces the result obtained earlier that

$$f_1 f_3 + f_2 \phi_3 = f_1 f_2 + k$$

is only reducible if $k = 0$.

It is also possible to express the relationship $ad - bc = 0$ in terms of the constants of equation (3.20), for the relationships (3.24) may be expressed in a determinant form in which the minors of A_3, B_3, C_3, and D_3 are a, b, c, and d. The condition becomes

$$\begin{vmatrix} a_1 & a_2 & a_3 \\ b_1 & b_2 & b_3 \\ c_1 & c_2 & c_3 \end{vmatrix} \times \begin{vmatrix} a_0 & a_1 & a_2 \\ b_0 & b_1 & b_2 \\ c_0 & c_1 & c_2 \end{vmatrix} - \begin{vmatrix} a_0 & a_2 & a_3 \\ b_0 & b_2 & b_3 \\ c_0 & c_2 & c_3 \end{vmatrix} \times \begin{vmatrix} a_0 & a_1 & a_3 \\ b_0 & b_1 & b_3 \\ c_0 & c_1 & c_3 \end{vmatrix} = 0.$$

These results can be summarized to give two canonical forms for the fourth order equation.

It has already been established that all equations, of the fourth order can be reduced to the form

$$f_1 f_3 + f_2 \phi_3 = f_1 f_2 + k.$$

If $k = 0$, then

$$\frac{f_3}{f_2} + \frac{\phi_3}{f_1} = 1$$

which is of the form

$$F_1 F_3 + F_2 \phi_3 = 1. \tag{3.26}$$

If $k \neq 0$, then a substitution of the form

$$f_2 = \frac{k}{F_2}$$

and other simple transformations lead to the form

$$F_1 F_2 F_3 + (F_1 + F_2)\, \phi_3 = 1. \tag{3.27}$$

Reduction to this last form is impossible if $k = 0$ showing that (3.26) and (3.27) are mutually exclusive.

Clark has therefore arrived at the important result that all fourth order equations can be reduced to one of the forms (3.26) or (3.27). Those which can be reduced to the type (3.26) can be represented by a nomogram of the fourth order, those reduced to the type (3.27) can not be represented by a nomogram of the fourth order. To the form (3.27) Clark gives the name *symmetric*.

Clark next proves a most important theorem which states that every equation of the third or fourth order is representable by a nomogram either of the same order or of two orders higher. The demonstration of this theorem is quite simple. Starting from equation (3.20) and dividing throughout by the term

$$c_0 f_1 f_2 + c_1 f_1 + c_2 f_2 + c_3$$

the resulting form is

$$x f_3 + y \phi_3 + \psi_3 = 0$$

in which

$$x = \frac{a_0 f_1 f_2 + a_1 f_1 + a_2 f_2 + a_3}{c_0 f_1 f_2 + c_1 f_1 + c_2 f_2 + c_3}$$

and

$$y = \frac{b_0 f_1 f_2 + b_1 f_1 + b_2 f_2 + b_3}{c_0 f_1 f_2 + c_1 f_1 + c_2 f_2 + c_3}.$$

From these last two relations it is possible to eliminate f_1 and f_2 in turn resulting in linear equations in x and y.

Three linear equations in x and y result,

$$x \chi_1 + y \phi_1 + \psi_1 = 0,$$
$$x \chi_2 + y \phi_2 + \psi_2 = 0, \tag{3.28}$$
$$\text{and} \quad x f_3 + y \phi_3 + \psi_3 = 0$$

from which it follows that

$$\begin{vmatrix} \chi_1 & \phi_1 & \psi_1 \\ \chi_1 & \phi_2 & \psi_2 \\ f_3 & \phi_3 & \psi_3 \end{vmatrix} = 0. \tag{3.29}$$

This shows that the variables of a fourth order equation can be separated into the determinant form. Furthermore, since this form can be written

$$\begin{vmatrix} G_1 & H_1 & 1 \\ G_2 & H_2 & 1 \\ G_3 & H_3 & 1 \end{vmatrix} = 0$$

it will represent, in general, an equation of the sixth order.

If in the original equation $\psi_3 = 0$, then this equation is of the third order. The determinant will now be of the form

$$\begin{vmatrix} G_1 & H_1 & 1 \\ G_2 & H_2 & 1 \\ G_3 & 0 & 1 \end{vmatrix} = 0$$

which, in general, is of the fifth order.

In this argument it is assumed that no linear relationship exists between G_1 and H_1 or between G_2 and H_2 since, if such a relationship did exist, then a normal type of nomogram would be possible.

A second remarkable theorem follows. It states that this method always leads to supports for the scales of α_1 and α_2 which are one and the same conic.

To see that this is so it is necessary first to demonstrate that α_1 and α_2 have supports which are conics. From the determinant form (3.29) it is seen that the supports of the scales are given by

$$x = \frac{\chi_1}{\psi_1} \quad \text{and} \quad y = \frac{\phi_1}{\psi_1} \quad \text{for } \alpha_1$$

and

$$x = \frac{\chi_2}{\psi_2} \quad \text{and} \quad y = \frac{\phi_2}{\psi_2} \quad \text{for } \alpha_2.$$

From the process by which the functions χ_i, ϕ_i, and ψ_i have been obtained it becomes clear after a simple investigation that, in general, the functions are of the second degree in f_i and consequently conics.

If $\chi_i \psi_i^{-1}$ is denoted by X_i and $\phi_i \psi_i^{-1}$ by Y_i then X_1 is a function of f_1 and X_2 is a function of f_2. To demonstrate that the conics are the same it is necessary to show that, if for some values of f_1 and f_2 we have $X_1 = X_2$, then $Y_1 = Y_2$.

The first two equations of (3.28) can be written as

$$xX_1 + yY_1 + 1 = 0$$
$$\text{and} \quad xX_2 + yY_2 + 1 = 0$$

and on subtraction the result is

$$x(X_1 - X_2) + y(Y_1 - Y_2) = 0$$

from which it immediately follows that if $X_1 = X_2$ then $Y_1 = Y_2$ unless x is infinite or y is zero, which will not generally be the case.

It is of interest to note that the inherent geometric symmetry is not reflected in an algebraic symmetry of the initial relationship as will be seen in the following example from Clark,

$$(1+l)h^2 - l(1+p)h - \frac{(1-l)(1+2p)}{3} = 0.$$

This is of nomographic order four since the four functions l, p, h, and h^2 are linearly independent.

It is an example of the form examined earlier when it was found that a nomogram of the fourth order could only be obtained if two of the factors happened to be identical. This is not the case here.

To apply Clark's method, divide throughout by one term, say $(1+l)$;

$$h^2 - l\frac{1+p}{1+l}h - \frac{(1-l)(1+2p)}{3(1+l)} = 0. \qquad (3.30)$$

Let

$$x = \frac{l(1+p)}{1+l} \quad \text{and} \quad y = \frac{(1-l)(1+2p)}{1+l} \qquad (3.31)$$

and (3.30) becomes

$$h^2 - xh - \frac{y}{3} = 0.$$

From equations (3.31), eliminate first p and then l, giving the two linear equations in x and y,

$$2x\frac{1+l}{l} - y\frac{1+l}{1-l} = 1$$

and

$$2x\frac{1}{1+p} + y\frac{1}{1+2p} = 1.$$

The determinant form obtained from these last three equations is

$$\begin{vmatrix} \frac{2(1+l)}{l} & -\frac{1+l}{1-l} & 1 \\ \frac{2}{1+p} & \frac{1}{1+2p} & 1 \\ h & \frac{1}{3} & h^2 \end{vmatrix} = 0.$$

The support of l is given by

$$x = \frac{2(1+l)}{l} \quad \text{and} \quad y = -\frac{1+l}{1-l}$$

whence, on eliminating l, $xy + x - 4y = 0$. The support of p is given by

$$x = \frac{2}{1+p}, \quad \text{and} \quad y = \frac{1}{1+2p}$$

which, on eliminating p, gives $xy + x - 4y = 0$. Thus l and p have the same conic as support.

One point needs to be emphasized. It is that if two scales have the same support the scales do not necessarily coincide. In general, every chord of the conic is a line of alignment.

The next logical step is to search for an equivalent symmetry in the algebraic relationship. Starting from the determinant form, if α_1 and α_2 are to have the same support then, if the support of α_1 is given by $x = f_1$ and $y = F(f_1)$ that of α_2 must be given by $x = f_2$ and $y = F(f_2)$ The determinant form must be

$$\begin{vmatrix} f_1 & F(f_1) & 1 \\ f_2 & F(f_2) & 1 \\ f_3 & \phi_3 & 1 \end{vmatrix} = 0.$$

If this is applied to the determinant form considered above, namely

$$\begin{vmatrix} \frac{2(1+l)}{l} & -\frac{1+l}{1-l} & 1 \\ \frac{2}{1+p} & \frac{1}{1+2p} & 1 \\ h & \frac{1}{3} & h^2 \end{vmatrix} = 0$$

then

$$f_1 = \frac{2(1+l)}{l} \quad \text{and} \quad f_2 = \frac{2}{1+p},$$

giving

$$l = \frac{2}{f_1 - 2} \quad \text{and} \quad p = \frac{2 - f_2}{f_2},$$

and

$$-\frac{1+l}{1-l} = \frac{f_1}{4 - f_1} \quad \text{and} \quad \frac{1}{1+2p} = \frac{f_2}{4 - f_2}$$

revealing the symmetry that was only implied. The revised determinant is

$$\begin{vmatrix} f_1 & \frac{f_1}{4-f_1} & 1 \\ f_2 & \frac{f_2}{4-f_2} & 1 \\ h & \frac{1}{3} & h^2 \end{vmatrix} = 0$$

showing explicitly the symmetry.

At this stage one can see that Clark's original aim has been achieved in that an equation, (3.30), of the fourth nomographic order for which no nomogram of the fourth order can be constructed, can nevertheless be represented by a nomogram associated with the sixth order, as illustrated in Figure 3.7.

It is necessary to bring together various pieces of Clark's work to see its overall significance. The following form represents a third order equation or a fourth order equation not representable by a chart of the same order. In

$$f_1 f_2 A_3 + (f_1 + f_2)B_3 + C_3 = 0 \tag{3.32}$$

let $x = f_1 f_2$ and $y = (f_1 + f_2)$ then, following Clark's procedure, the following linear equations are obtained:

$$x - yf_1 + f_1^2 = 0,$$
$$x - yf_2 + f_2^2 = 0,$$
$$\text{and} \quad xA_3 + yB_3 + C_3 = 0.$$

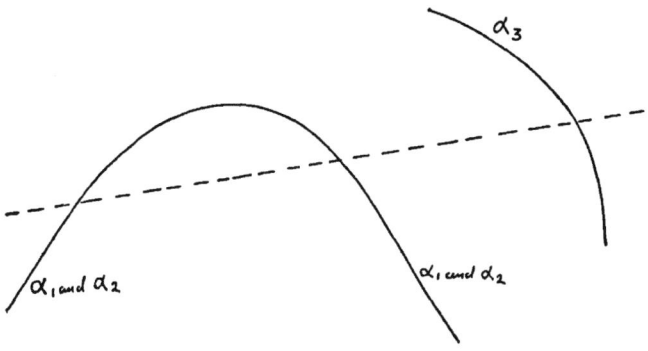

FIGURE 3.7

These lead to the determinant

$$\begin{vmatrix} f_1 & f_1^2 & 1, \\ f_2 & f_2^2 & 1 \\ -B_3 & C_3 & A_3 \end{vmatrix} = 0 \tag{3.33}$$

from which it is seen that the support of α_1 and of α_2 is the parabola $y = x^2$.

A homographic transformation can always transform this parabola into another conic, a circle having obvious advantages.

To the determinant (3.33) and its associated equation (3.32) Clark gives the name *canonical form of conical charts* because it encompasses all equations to which the method applies; i.e., all equations of the third order and all those of the fourth order not representable by a chart of the same order.

It is of interest to look at some examples of this form. Consider the multiplication formula $f_1 f_2 = f_3$ which led to one of the earliest nomograms of Lalanne. It is obtained by putting $A_3 = 1$, $B_3 = 0$, and $C_3 = -f_3$ in (3.32) giving the determinant

$$\begin{vmatrix} f_1 & f_1^2 & 1 \\ f_2 & f_2^2 & 1 \\ 0 & -f_3 & 1 \end{vmatrix} = 0$$

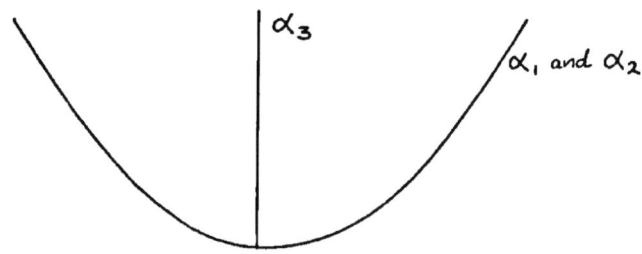

FIGURE 3.8. $f_1 f_2 = f_3$

which has as support for α_1 and α_2 and the parabola $y = x^2$ and for α_3 the line $x = 0$ which is the axis of that parabola, (Figure 3.8).

For addition, $f_1 + f_2 = f_3$, we have $A_3 = 0$, $B_3 = 1$, and $C_3 = -f_3$ so the determinate is

$$\begin{vmatrix} f_1 & f_1^2 & 1 \\ f_2 & f_2^2 & 1 \\ -1 & -f_3 & 0 \end{vmatrix} = 0$$

which needs to be transformed by dividing by the terms in the middle column giving

$$\begin{vmatrix} f_1^{-1} & f_1^{-2} & 1 \\ f_2^{-1} & f_2^{-2} & 1 \\ f_3^{-1} & 0 & 1 \end{vmatrix} = 0.$$

Again α_1 and of α_2 lie on the parabola $y = x^2$ but α_3 now lies on $y = 0$, the tangent to the vertex, (Figure 3.9).

The two preceding examples are capable of being represented by straight line nomograms; i.e., of order 3, but here their representation has been of order 5. The following example has a negative discriminant and therefore cannot be represented by straight lines. Begin by noting that

$$\tan(\alpha + \beta) = \frac{\tan \alpha + \tan \beta}{1 - \tan \alpha \tan \beta}$$

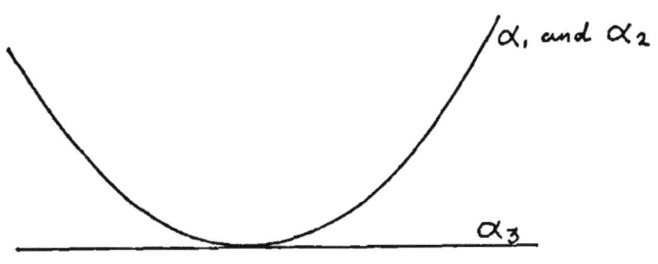

FIGURE 3.9. $f_1 + f_2 = f_3$

can be written as

$$f_3 = \frac{f_1 + f_2}{1 - f_1 f_2}$$

or

$$f_1 f_2 f_3 + f_1 + f_2 - f_3 = 0.$$

The determinant is

$$\begin{vmatrix} f_1 & f_1^2 & 1 \\ f_2 & f_2^2 & 1 \\ -1 & -f_3 & f_3 \end{vmatrix} = 0$$

or

$$\begin{vmatrix} f_1 & f_1^2 & 1 \\ f_2 & f_2^2 & 1 \\ -f_3^{-1} & -1 & 1 \end{vmatrix} = 0$$

in which case α and β lie on the parabola $y = x^2$ while $\alpha + \beta$ is on the line $y = -1$, (Figure 3.10).

These three examples all illustrate third order equations and it is interesting to note in passing that the method offers a flexibility, in addition to that available as a result of homographic transformation, which is not possible in the case of fourth order equations which yield to this method. The flexibility lies in the ability to modify the relative positions of the scales.

In the case of $f_1 f_2 = f_3$, the equation can be rewritten as $f_1(\lambda f_2) = \lambda f_3$, which will give symmetry with respect to f_1 and λf_2 . Similarly, $f_1 + f_2 = f_3$ can be written as $f_1 + (f_2 + \lambda) = (f_3 + \lambda)$, which is symmetric in f_1 and $f_2 + \lambda$.

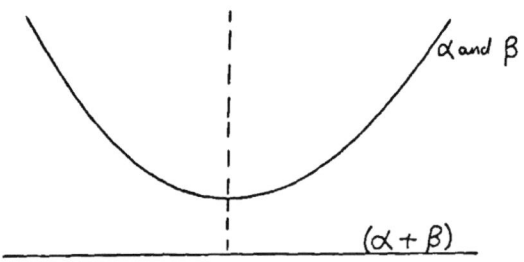

FIGURE 3.10. $\tan(\alpha + \beta) = \frac{\tan\alpha + \tan\beta}{1 - \tan\alpha\tan\beta}$

The value of this device is seen in the case of $f_1 f_2 = f_3$ if it is written $f_1(-f_2) = -f_3$ giving the determinant

$$\begin{vmatrix} f_1 & f_1^2 & 1 \\ -f_2 & f_2^2 & 1 \\ 0 & f_3 & 1 \end{vmatrix} = 0.$$

In this version the scales of f_1 and f_2 are in opposite directions from the axis, (Figure 3.11).

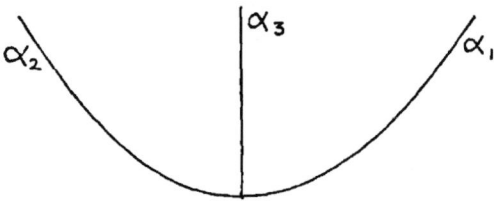

FIGURE 3.11. $f_1(-f_2) = -f_3$

Clark gives construction methods for his conical nomograms. He calls this assembly of techniques the *synthetic method* for the construction of conical charts. Firstly, he describes a method of d'Ocagne's for the construction of a linear scale in which to fix a scale it is only necessary to know three points on it, $\alpha = \infty$ and two others, say, $\alpha = 0$ and $\alpha = 1$. He then uses this to solve the analogous problem of constructing a linear sheaf of rays from three dimensional rays. There is a simple geometric construction for this.

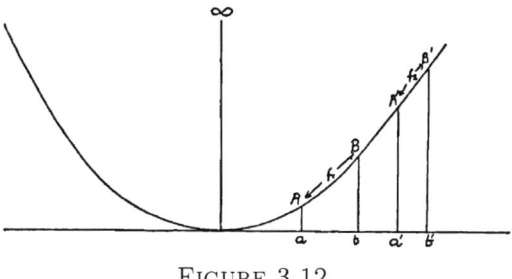

FIGURE 3.12

To construct the nomogram, Clark starts with the determinant form

$$\begin{vmatrix} f_1 & f_1^2 & 1 \\ -f_2 & f_2^2 & 1 \\ -B_3 & C_3 & A_3 \end{vmatrix} = 0.$$

In practice f_1 and f_2 will have limits to their ranges. In Figure 3.12 the limits are a and b for f_1 and a' and b' for f_2.

The anharmonic ratio $(AA'BB') = k$ is a fundamental property of the arrangement.

If the scales AB and $A'B'$ are required to be opposite each other, and $AA'BB'$ is to form a rectangle, then a suitable rectangle $AA'BB'$ is drawn and the vertices a, b, a', and b' are marked as in Figure 3.13. Then, taking A' as the origin of a sheaf the three rays $A'B$, $A'A$, and $A'B'$ are sufficient to determine the sheaf.

Similarly, B' can be taken as the origin of three rays $B'A'$, $B'B$, and $B'A$ and this sheaf completed. The intersection of the corresponding rays will then give the scale of f_1. A similar construction from A and B will give the scale of f_2.

The conic is therefore determined by the choice of four dimensioned points A, B, A', and B'. The order in which the points are positioned determines whether the conic is an ellipse or a hyperbola. Figures 3.14 and 3.15 illustrate this.

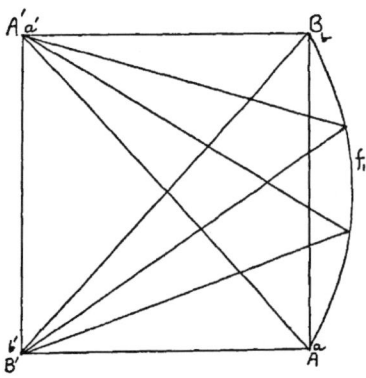

FIGURE 3.13

As for the third scale, if it is exterior to the canonical parabola it will also be exterior to any other conic chosen.

The alternative to choosing the four points is to choose the conic. In this case the anharmonic ratio k must be preserved. For the circle of Figure 3.16 this results in the angle α being determined by $\tan^2 \alpha = k$.

To illustrate these points, consider equation (3.30) for which the determinant form is

$$\begin{vmatrix} \frac{2(1+l)}{l} & -\frac{1+l}{1-l} & 1 \\ \frac{2}{1+p} & \frac{1}{1+2p} & 1 \\ h & \frac{1}{3} & h^2 \end{vmatrix} = 0.$$

The relationship between l and p is given by equating the first terms of the first two rows, giving

$$l = -\frac{1+p}{p} \quad \text{or} \quad p = -\frac{1}{1+l}. \tag{3.34}$$

Suppose that the extreme values of the scales l and p are $\frac{1}{2}$ and 1 in both cases. The corresponding values of l and p are found from (3.34) and are given in Table 3.2. They are used to construct the scales of p and l as indicated in Figures 3.17 and 3.18.

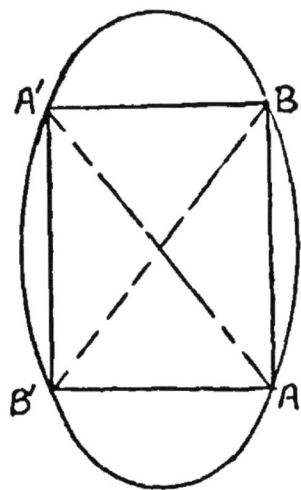

FIGURE 3.14

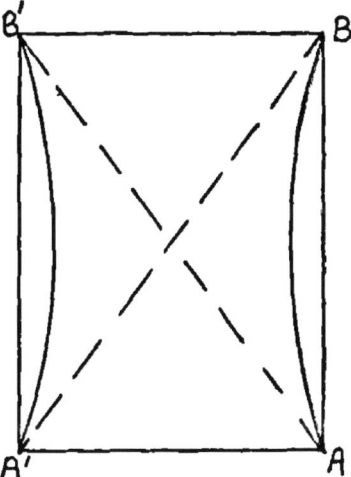

FIGURE 3.15

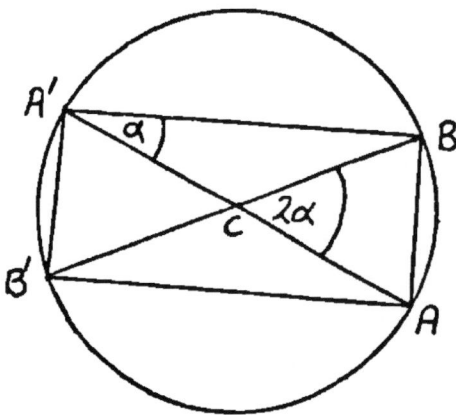

FIGURE 3.16

The h scale can be found by locating four values of h. For example, if $h = 0$ the original equation gives $l = 1$ and $p = -\frac{1}{2}$ which locates $h = 0$ at A. Similarly $h = 1$ is at A'.

Points	A	B	A'	B'
p	1	$\frac{1}{2}$	$-\frac{1}{2}$	$-\frac{2}{3}$
l	-2	-3	1	$\frac{1}{2}$

TABLE 3.2.

To find the value of h at which its support cuts AB', put $p = 1$ and $l = \frac{1}{2}$ in the original equation; i.e., make use of the alignment of p, l, and h. This gives $h = -\frac{1}{3}$ (or 1 which has already been found). The point for which $h = -\frac{1}{3}$ lies somewhere on AB'. To fix it, put $h = -\frac{1}{3}$ and $p = \frac{1}{2}$ (the point B) in the original equation; this gives $l = \frac{10}{23}$ and enables the point C, where $h = -\frac{1}{3}$ to be fixed precisely. Similarly, the point D on $A'B$ where $h = \frac{3}{4}$ is located. The h scale can now be completed using the sheaves emanating from B and B'.

Figure 3.19 illustrates the arrangement of the scales. Figure 3.20 showing the construction lines, is reproduced from Clark's paper.

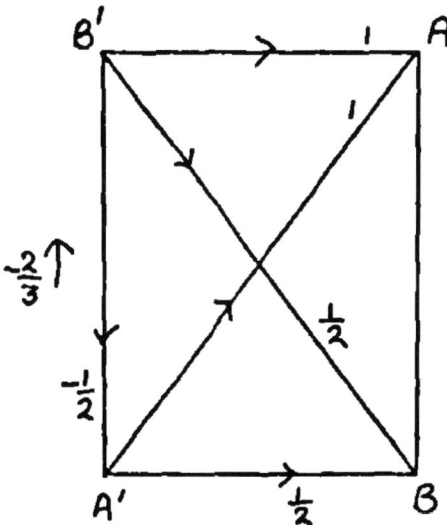

FIGURE 3.17. Sheaf for the construction of p

In his synthetic method Clark has achieved a transformation from one conic to another by a simple geometric procedure. In order to emphasize this simplicity he illustrates the complexity of the analytical transformation of a parabola $y = x^2$ and its axis $x = 0$ into a circle and a chord of that circle. There is nothing new about his method but his point is well made. However, he concludes that the real advantage of the synthetic method is not so much to avoid a complicated analysis as to be able to start with the most advantageous disposition of the scales and to know, a priori, the degree of freedom that one has.

In the development so far one fact has not been brought out. It is that in Clark's method for getting the determinant from $f_1 f_2 A_3 + (f_1 + f_2) B_3 + C_3 = 0$; i.e., by letting $x = f_1 f_2$ and $y = f_1 + f_2$, an extraneous factor is introduced. This can be seen from the determinant form,

$$\begin{vmatrix} f_1 & f_1^2 & 1 \\ f_2 & f_2^2 & 1 \\ -B_3 & C_3 & A_3 \end{vmatrix} = 0$$

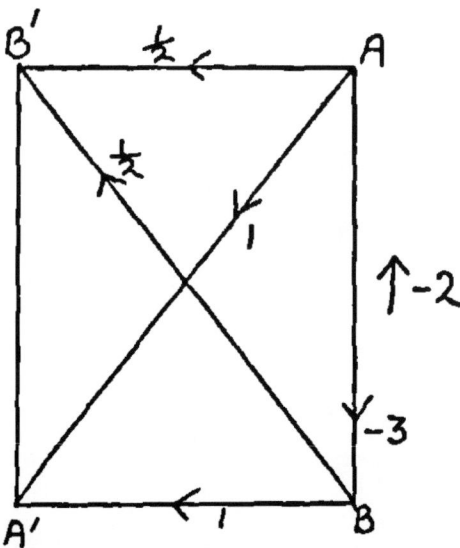

FIGURE 3.18. Sheaf for the construction of l

either by noting that if $f_1 = f_2$ it becomes identically zero and therefore $(f_1 - f_2)$ must be a factor, or by expanding it. The expanded determinant gives,

$$(f_2 - f_1)(f_1 f_2 A_3 + (f_2 + f_1)B_3 + C_3) = 0.$$

Clark then poses this question. If it is possible to achieve symmetry in two variables through multiplication by an extraneous factor, then is it possible that symmetry in three variables can be achieved through multiplication by some other extraneous factor?

To develop this line of thought Clark considers an equation of the third nomographic order put into a symmetric form with respect to the three variables (or three functions of these variables). The form is,

$$f_1 f_2 f_3 + A(f_1 f_2 + f_2 f_3 + f_1 f_3) + B(f_1 + f_2 + f_3) + C = 0. \qquad (3.35)$$

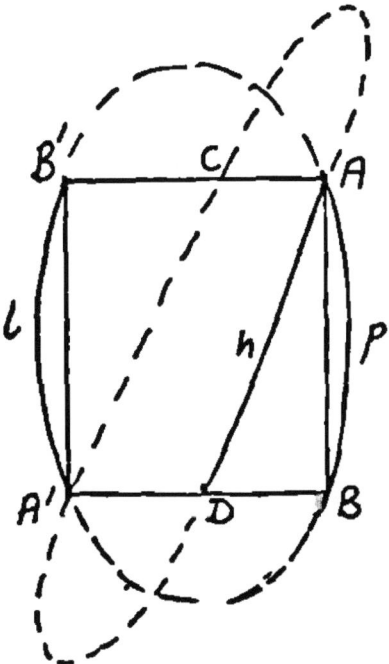

<small>FIGURE 3.19</small>

Symmetry of three rows of a determinant in f_1, f_2, and f_3 and the previous result suggests an extraneous factor of

$$S = (f_1 - f_2)(f_2 - f_3)(f_3 - f_1).$$

In determinant form, S is given by

$$\begin{vmatrix} f_1 & f_1^2 & 1 \\ f_2 & f_2^2 & 1 \\ f_3 & f_3^2 & 1 \end{vmatrix}$$

Equation (3.35) multiplied by S gives

$$Sf_1f_2f_3 + SA(f_1f_2 + f_2f_3 + f_1f_3) + \tag{3.36}$$
$$SB(f_1 + f_2 + f_3) + SC = 0$$

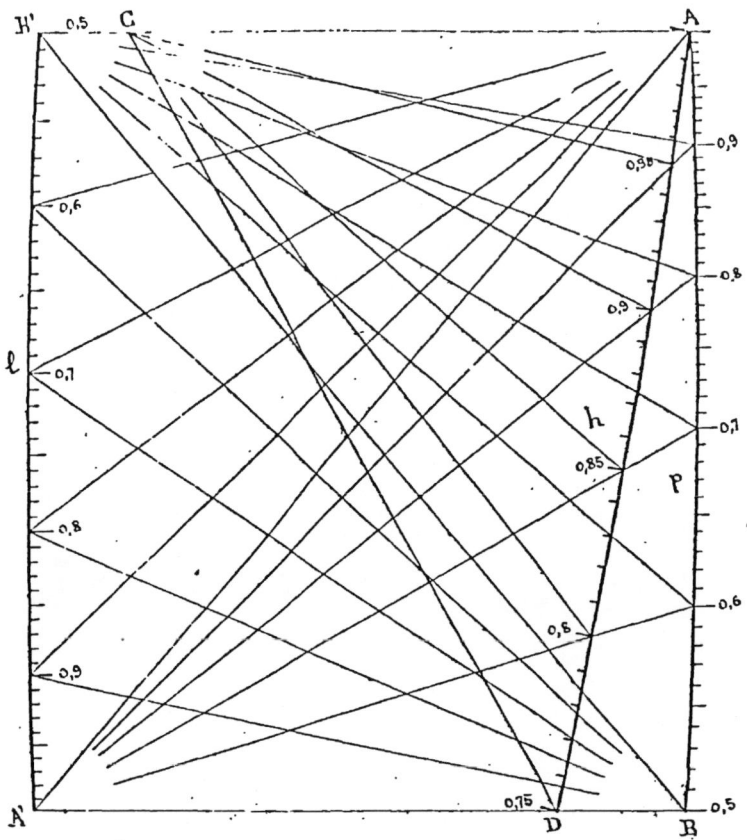

FIGURE 3.20. Facsimile of Clark's nomogram for $(1 + l)h^2 - l(1 + p)h - \frac{(1-l)(1+2p)}{3} = 0$ showing the construction lines. The drawing is the work of G.Fleuri, the translator of Clark's paper into French

which has the determinant form

$$
\begin{vmatrix}
1 & -A & B & -C \\
1 & f_1 & f_1^2 & f_1^3 \\
1 & f_2 & f_2^2 & f_2^3 \\
1 & f_3 & f_3^2 & f_3^3
\end{vmatrix} = 0.
\qquad (3.37)
$$

Since (3.37) is the determinant form of (3.36) it is useful to note that the determinant forms of $Sf_1f_2f_3$, $S(f_1 + f_2 + f_3)$, and $S(f_1f_2 + f_2f_3 + f_1f_3)$ are contained in (3.37) as the minors of the elements of the first row.

For example,

$$S(f_1 + f_2 + f_3) = \begin{vmatrix} 1 & f_1 & f_1^3 \\ 1 & f_2 & f_2^3 \\ 1 & f_3 & f_3^3 \end{vmatrix}.$$

If the row operations

$$\text{row}_2 - \text{row}_1,$$
$$\text{row}_3 - \text{row}_1,$$
$$\text{row}_4 - \text{row}_1$$

are performed on (3.37), the determinant becomes

$$\begin{vmatrix} 1 & -A & B & -C \\ 0 & f_1 + A & f_1^2 - B & f_1^3 + C \\ 0 & f_2 + A & f_2^2 - B & f_2^3 + C \\ 0 & f_3 + A & f_3^2 - B & f_3^3 + C \end{vmatrix} = 0$$

which can be expressed as the third order determinant

$$\begin{vmatrix} f_1 + A & f_1^2 - B & f_1^3 + C \\ f_2 + A & f_2^2 - B & f_2^3 + C \\ f_3 + A & f_3^2 - B & f_3^3 + C \end{vmatrix} = 0. \tag{3.38}$$

This shows that the supports of f_1, f_2, and f_3 are one and the same curve. The curve is given by the parametric equations

$$x = \frac{f_1 + A}{f_1^3 + C} \quad \text{and} \quad y = \frac{f_1^2 - B}{f_1^3 + C}$$

which is of the form

$$y^3 + x^3C = yx + Ax^2 \tag{3.39}$$

$$\text{or} \quad y^3 + x^3C = yx - B\frac{x^3}{y}.$$

However, more important than the exact form of the equation is the fact that the common support of all three scales is a curve of the third degree. Consequently, the resulting nomograms are called *cubic*.

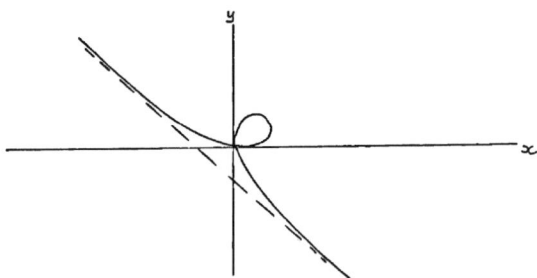

FIGURE 3.21. The Folium of Descartes

As an example consider the multiplication formula $f_1 f_2 = \phi_3$ which Lalanne used to illustrate anamorphosis and which has been used as an illustration of Clark's conical nomogram.

If in $f_1 f_2 = \phi_3$, ϕ_3 is replaced by f_3^{-1} the relationship $f_1 f_2 f_3 = 1$ is obtained. It is also obtained from (3.35) by making $A = B = 0$ and $C = -1$, showing that the common curve is $y^3 - x^3 = yx$. However, the curve $y^3 + x^3 = yx$ is also acceptable since the determinant equation remains true if any column is multiplied by -1; thus in this case y has been replaced by $-y$.

As an example Clark has used the form $y^3 + x^3 = axy$. Here, in the determinant form

$$\begin{vmatrix} f_1 & f_1^2 & f_1^3 - 1 \\ f_2 & f_2^2 & f_2^3 - 1 \\ f_3 & f_3^2 & f_3^3 - 1 \end{vmatrix} = 0,$$

he has chosen to multiply column$_1$ by a and column$_2$ by -1. He expected a lot from his readers for he gives no explanation of how he has moved from the previous determinant to

$$x = a\frac{f}{f^3 - 1} \quad \text{and} \quad y = -a\frac{f^2}{f^3 - 1}.$$

The curve $x^3 + y^3 = axy$ is the folium of Descartes and has the general appearance shown in Figure 3.21.

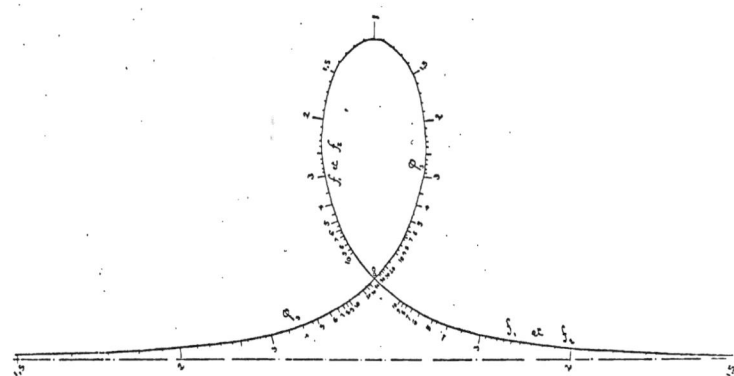

FIGURE 3.22. Clark's cubic nomogram for multiplication: $f_1 f_2 = \phi_3$

Marking the curve with the values of the variables is made easier by noting the linear relationship $\frac{y}{x} = -f$. The nomogram from Clark's paper is given as Figure 3.22.

For a second example consider

$$f_1 + f_2 + f_3 = 0.$$

The determinant form for this is best obtained by considering (3.36) and taking the appropriate minor from (3.37). It is easily seen to be that

$$\begin{vmatrix} 1 & f_1 & f_1^3 \\ 1 & f_2 & f_2^3 \\ 1 & f_3 & f_3^3 \end{vmatrix} = 0.$$

Clark's method is to make the substitution in (3.38) of $A = C = 0$ and $B = \infty$ which, if the second column is divided by B and multiplied by -1 before the substitution is made, does give the determinant but it seems inelegant in view of the algebraic form of (3.36). The curve is given by $x = f$, $y = f^3$ or $y = x^3$, which is a cubic parabola as sketched in Figure 3.23.

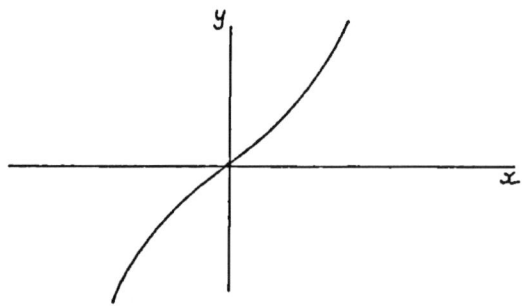

FIGURE 3.23. A cubic parabola

For the relationship $f_1^{-1} + f_2^{-1} + f_3^{-1} = 0$, note that it can be also written as

$$(f_1 f_2 f_3)^{-1}(f_2 f_3 + f_1 f_3 + f_1 f_2) = 0.$$

The determinant form for $f_2 f_3 + f_1 f_3 + f_1 f_2$ is the minor of A in (3.36); i.e.,

$$\begin{vmatrix} 1 & f_1^2 & f_1^3 \\ 1 & f_2^2 & f_2^3 \\ 1 & f_3^2 & f_3^3 \end{vmatrix} = 0.$$

This will serve as the determinant since if it is multiplied by $(f_1 f_2 f_3)^{-1}$ it becomes

$$\begin{vmatrix} f_1^{-1} & f_1 & f_1^2 \\ f_2^{-1} & f_2 & f_2^2 \\ f_3^{-1} & f_3 & f_3^2 \end{vmatrix} = 0$$

which reverts to

$$\begin{vmatrix} 1 & f_1^2 & f_1^3 \\ 1 & f_2^2 & f_2^3 \\ 1 & f_3^2 & f_3^3 \end{vmatrix} = 0$$

on dividing the elements of each row by the corresponding element in the first row.

The curve is given by $x = f^2$ and $y = f^3$; i.e., $x^3 = y^2$, as illustrated in Figure 3.24.

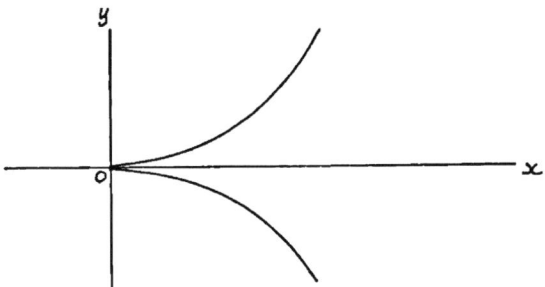

FIGURE 3.24. A cubic parabola

As a third example consider

$$\tan(\alpha + \beta) = \frac{\tan \alpha + \tan \beta}{1 - \tan \alpha \tan \beta}.$$

If

$$\tan \alpha = f_1, \quad \tan \beta = f_2, \quad \text{and} \quad \tan(\alpha + \beta) = -f_3,$$

then the relationship takes the form

$$f_1 f_2 f_3 - (f_1 + f_2 + f_3) = 0.$$

Comparing this with equation (3.35), $A = 0$, $B = -1$, and $C = 0$ so determinant (3.38) becomes

$$\begin{vmatrix} f_1 & f_1^2 + 1 & f_1^3 \\ f_2 & f_2^2 + 1 & f_2^3 \\ f_3 & f_3^2 + 1 & f_3^3 \end{vmatrix} = 0.$$

On adding the first column to the third and dividing by the corresponding element of the second column the determinant becomes

$$\begin{vmatrix} 1 & f_1 & \dfrac{f_1}{f_1^2 + 1} \\[2ex] 1 & f_2 & \dfrac{f_2}{f_2^2 + 1} \\[2ex] 1 & f_3 & \dfrac{f_3}{f_3^2 + 1} \end{vmatrix} = 0.$$

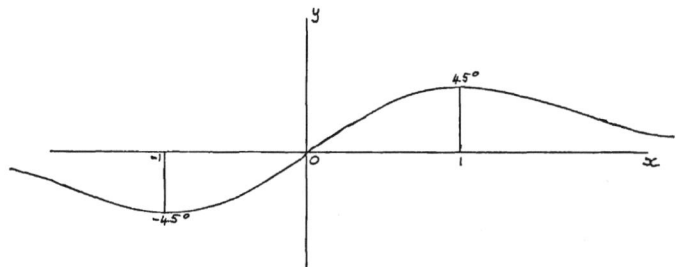

FIGURE 3.25. $y = x(x^2 + 1)^{-1}$

The curve is given by $x = f$ and

$$y = \frac{f}{f^2 + 1} \quad \text{or} \quad y = \frac{x}{x^2 + 1}$$

which is of the type shown in Figure 3.25.

To position the scales on the curve note that $x = f$, a regular scale on the x axis. It then only requires parallels to the y axis to be drawn through the points of this scale to intersect the curve, to give the scale. The example of this given by Clark is reproduced as Figure 3.26.   As in the case of conical nomograms, the scales on cubic nomograms can be displaced relative to each other by the introduction of two constant factors. For example, $f_1 f_2 f_3 = 1$ can be written as $f_1(\lambda f_2)(\mu f_3) = \lambda\mu$ giving alignment in which $f_1 = \lambda f_2 = \mu f_3$.

The cubic nomograms are curves of the type known as *cubic unicursal*. A unicursal curve is one with parametric equations $x = \theta(t)$ and $y = \phi(t)$ in which θ and ϕ and are rational functions of t. Such a curve cannot consist of, for example, an ellipse and a separated branch. They are quite different from conics and it is not possible for a cubic nomogram to degenerate into a conical one. The same type of distinction can be made between conical nomograms and ordinary alignment ones and it follows that the three types, ordinary, conic and cubic must be regarded as quite distinct types of alignment nomogram.

It is of interest to ask how many different types of cubic unicursal are possible; that is, how many types are there which cannot be obtained by projective transformations of other types? To decide this question it is necessary to have

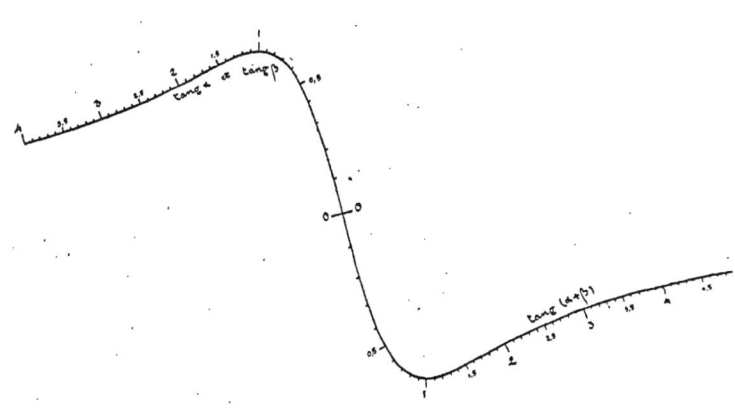

FIGURE 3.26. Clark's cubic nomogram for $\tan(\alpha + \beta) = \frac{\tan\alpha + \tan\beta}{1 - \tan\alpha\tan\beta}$

some means of classification of cubic unicursals. A classification system is based on the double point that all cubic unicursals must possess. The two tangents at this point may be real and distinct, or real and coincident or imaginary and it is this distinction in the nature of the tangents that is used to classify cubic unicursals.

The folium of Descartes given by $x^3 + y^3 = xy$ has a double point at the origin. The tangents at this point are real and distinct and are given by $x = 0$ and $y = 0$. This is known as the *crunodal form* (Figure 3.27). The form $x^3 = y^2$ has coincident tangents at the origin given by $y = 0$. This form is known as the *cuspidal form* (Figure 3.28). The form

$$y = \frac{x}{x^2 + 1}$$

has an isolated point, which is a double point, at the origin and therefore has two imaginary tangents at that point. This is known as the *acnodal form* (Figure 3.29).

These three forms are quite distinct and cannot be transformed one into another by projection. In this respect they differ from the conical nomograms in which any conic can be projected into any other conic.

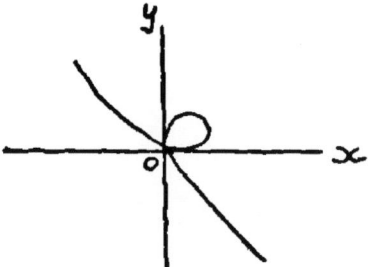

FIGURE 3.27

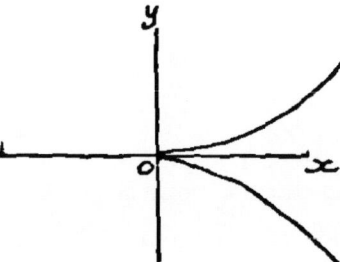

FIGURE 3.28

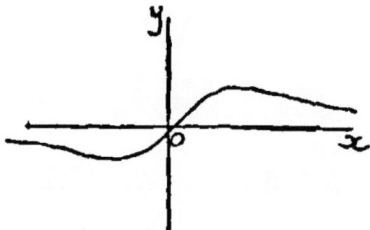

FIGURE 3.29

The relationships which lead to the three cubic forms were,

  i. $f_1 f_2 f_3 = 1$ for the crunodal form
  ii. $\phi_1 + \phi_2 + \phi_3 = 0$ (where $\phi_i = f_i^{-1}$) for the cuspidal form.
  iii. $f_1 f_2 f_3 - (f_1 + f_2 + f_3) = 0$ for the acnodal form.

In the context of ordinary alignment nomograms (i) is the type with non-concurrent linear scales, (ii) the type with concurrent linear scales and (iii) is Clark's irreducible form.

This correspondence between the cubic and ordinary alignment nomograms is based on their projective properties; one type cannot be projected into another.

In a section headed "Conclusion of the theory of ample alignment charts. General Method", Clark brings together the conclusions of his investigations. He addresses himself to the fundamental question which had been posed at the beginning of his work, namely, the possibility of representing a given equation

$$f(\alpha_1, \alpha_2, \alpha_3) = 0$$

by a nomogram. It is perhaps kind to pass over without comment his remark that "The simplicity of this problem can now be shown in its true light".

The essence of his method is to bring together the functions of one variable, say α_3 such that the expression is linear in the functions of that variable, say f_3, ϕ_3, and ψ_3 giving

$$f_3 A_{12} + \phi_3 B_{12} + \psi_3 C_{12} = 0.$$

Then, putting

$$x = \frac{A_{12}}{C_{12}} \quad \text{and} \quad y = \frac{B_{12}}{C_{12}},$$

a linear equation in α_3 is obtained; viz.

$$x f_3 + y \phi_3 + \psi_3 = 0.$$

If the expressions for x and y, when α_1 and α_2 are eliminated in turn, yield equations linear in x and y, the problem is solved for suppose that they give

$$x f_2 + y \phi_2 + \psi_2 = 0$$

and
$$xf_1 + y\phi_1 + \psi_3 = 0,$$
then the determinant
$$\begin{vmatrix} f_1 & \phi_1 & \psi_1 \\ f_2 & \phi_2 & \psi_2 \\ f_3 & \phi_3 & \psi_3 \end{vmatrix} = 0$$
holds.

If linear equations in x and y are not obtained Clark says that the expression is irreducible. There is much to be said on this problem later but at this stage two observations need to be made. The first is that in the form given above it is a very useful contribution. The second is that the concept is not new. As d'Ocagne points out in his introduction he had given the same idea in his *Nomographie* of 1891 [**90**], but neither d'Ocagne nor Clark acknowledge the true originator of the method, Massau, who gave the method in 1884 [**82**]. Massau had also observed that the method could produce an extraneous factor.

It had originally been Clark's intention to produce canonical forms for equations of the fifth and sixth nomographic orders but he later deemed it unnecessary, considering that the work already done would embrace these types. He gives two examples. Firstly a symmetric equation of the sixth order,
$$f_3(f_1^2 + f_2^2 + f_1 f_2) - \phi_3(f_1^2 f_2 + f_1 f_2^2) = 1$$
in which he puts
$$x = f_1^2 + f_2^2 + f_1 f_2 \quad \text{and} \quad y = f_1^2 f_2 + f_1 f_2^2$$
finally arriving at the system of equations,
$$xf_1 - y = f_1^3,$$
$$xf_2 - y = f_2^3,$$
$$\text{and} \quad xf_3 - y\phi_3 = 1$$
which leads to
$$\begin{vmatrix} f_1 & f_1^3 & 1 \\ f_2 & f_2^3 & 1 \\ f_3 & 1 & \phi_3 \end{vmatrix} = 0.$$
Secondly a non-symmetric equation of the sixth order,
$$f_3(f_1^2 - f_2^3) - \phi_3 f_1 f_2(f_1 - f_2^2) = f_1 - f_2$$

which, treated in the same way, yields the determinant,

$$\begin{vmatrix} f_1 & f_1^2 & 1 \\ f_2 & f_2^3 & 1 \\ f_3 & 1 & \phi_3 \end{vmatrix} = 0.$$

In the non-symmetric example the expanded determinant form gives precisely the equation from which it has been derived, whereas in the symmetric example the determinant contains an extraneous factor $(f_1 - f_2)$.

Towards the end of his paper Clark draws some conclusions arrived at as a result of his development of the material for the paper. He considers the notion of nomographic order to be no more than a useful preliminary criterion of representation. The problem of expressing a given equation

$$F(\alpha_1, \alpha_2, \alpha_3) = 0$$

as the eliminant of three linear equations he considers to be the true problem. Then, considering the determinant he recognizes the following two types:

> Non-symmetric: exactly representing the given equation
> Symmetric: representing the given equation multiplied by an extraneous factor.

In type (i) the supports of the scales are distinct. In type (ii) some supports will be coincident. The extraneous factor equated to zero will indicate the supports which coincide. The extraneous factor must contain either (a) two of the variables α_1, α_2, and α_3 when the factor is of the form $(f_i - f_j)$ as in the conical nomograms, or (b) the three variables α_1, α_2, and α_3, when the factor is of the form $(f_1 - f_2)(f_2 - f_3)(f_3 - f_1)$.

An extraneous factor in the determinant always implies symmetry and coincidence of scales while no extraneous factor implies no coincidence of scales. To demonstrate this further he extends an argument already used. If the elimination process results in the two equations being of the type

$$xx_1 + yy_1 + 1 = 0 \quad \text{and} \quad xx_2 + yy_2 + 1 = 0,$$

then subtraction gives

$$x(x_1 - x_2) + y(y_1 - y_2) = 0.$$

If α_1 and α_2 have values such that $x_1 = x_2$, then it may be that $y_1 = y_2$, in which case the supports coincide and the functions determined by the values of x and y, namely f_1 and f_2, give the extraneous factor $(f_1 - f_2)$.

However, another possibility, given that $x_1 = x_2$, is that either $x = \infty$ or $y = 0$ as can be seen if the expression is written as

$$(x_1 - x_2) = \frac{y}{x}(y_2 - y_1).$$

In this case the supports are distinct and no common factor can be found.

Clark is now able to rephrase the real problem as that of finding in which cases the left hand side of

$$F(\alpha_1, \alpha_2, \alpha_3) = 0$$

can be represented as a determinant either as it is or after multiplication by a factor.

He is of the opinion that the elimination method solves the problem in all cases except where the required factor contains all three variables, in which cases if a nomogram is possible, it must be of the particular form associated with a cubic nomogram.

This claim of Clark's must be taken in the spirit in which it is made. Firstly, he has only concerned himself with the algebraic problem of representation and recognizes that more powerful methods than his own, or Saint-Robert's or Massau's, will be required to "settle the complete representation problem." Secondly, he has an attitude of mind which is revealed in his introduction, where, referring to the elimination process, he states that the method gives a nomogram or does not, according as that representation is possible or not. This is the constructionist approach which was put forward much later, in relation to nomograms, by the Russian Džems-Levi. It seems that Clark's philosophy of mathematics was ahead of its time.

There is a final section to Clark's paper which applies his method of conical charts to equations having four variables. The general approach is not new for it involves the introduction of an auxiliary variable. As an example, consider $F(\alpha_1, \alpha_2, \alpha_3, \alpha_4) = 0$. This is rewritten

$$f_1(\alpha_1, \alpha_2, \alpha) = 0 \quad \text{and} \quad f_2(\alpha_3, \alpha_4, \alpha) = 0.$$

Then, effectively, two nomograms are constructed having the common scale α. Clark extends this method by making his scale α coincide with that of α_1 in f_1 and α_3 in f_3 for example. If the conic carrying the coincident scales is a circle the appearance of the chart can be greatly simplified. However, it is the application which is new and it will be sufficient here to note it.

3.5. Further Theoretical Developments of Alignment Nomograms

During the period between the time when Clark made public his ideas at Cherbourg and the appearance of those ideas in print, d'Ocagne published four papers which, in one way or another, were related to the unpublished work of Clark.

In 1906 he showed that Clark's conical nomograms had a link with his own alignment nomograms with three rectilinear scales [**97**]. This link is quite easy to show. Using the notation that d'Ocagne had used in expounding his theory, (3.5) et seq., the general equation in the three variables α_1, α_2, and α_3 is written,

$$Af_1f_2f_3 + B_1f_2f_3 + B_2f_1f_3 + B_3f_1f_2+ \tag{3.40}$$
$$C_1f_1 + C_2f_2 + C_3f_3 + D = 0.$$

If u and v are parallel coordinates and the substitutions

$$u = Af_1f_2 + B_1f_2 + B_2f_1 + C_3 \tag{3.41}$$
$$\text{and} \quad v = B_3f_1f_2 + C_1f_1 + C_2f_2 + D$$

are made in (3.40), the result is $uf_3 + v = D$, showing that the variable α_3 has for support the straight line which is the axis of the origins.

If from (3.41), f_1 is eliminated the result is,

$$E_2f_2^2 + (B_2u - Av - F_2)f_2 + C_1u - B_2v + C_2 = 0.$$

If f_2 is eliminated instead, the result is,

$$E_1f_1^2 + (B_2u - Av - F_1)f_2 + C_2u - B_1v + C_1 = 0.$$

These two results show that both α_1 and α_2 have conics for supports. It remains now to show that these conics coincide. To do this he finds the support of α_2 which is

$$(B_3u - Av - F_2)^2 - 4E_2(C_1u - B_2v + C_2) = 0,$$

and he then investigates whether the index 2 is fundamental to this expression. He rewrites the expression as

$$(B_3u - Av)^2 - 2(F_2B_3 + 2E_2C_1)\,u +$$
$$2(AF_2 + 2E_2B_2)v + \Delta = 0.$$

It is clear that the first and last terms are independent of the index 2. The other two terms may be written as follows:

$$F_2 B_3 + 2E_2 C_1 = B_3(F_0 - 2B_1 C_1 - 2B_2 C_2) + 2AC_1 C_2$$

and

$$AF_2 + 2E_2 B_2 = AF_0 - 2B_1 B_2 B_3$$

neither of which change their values if the indices 1 and 2 are interchanged, thus α_1 and α_2 have the same supports. The impact of this result is somewhat diminished by the knowledge that the idea was originally Clark's. However, the academic interest of it is considerable.

Some nine months later d'Ocagne's paper on critical points appeared [99]. I have referred to this notion already, indicating that Massau was the originator of the idea and that this fact is nowhere mentioned by d'Ocagne. D'Ocagne developed the concept beyond the point reached by Massau and was able to graft it onto his earlier theory.

His treatment again begins with the general third order equation (3.40) which for this purpose is rewritten,

$$f_1(Af_2 f_3 + B_2 f_3 + B_3 f_2 + C_1) + B_1 f_2 f_3 + C_2 f_2 + C_3 f_3 + D = 0$$

It can be seen that f_1 is indeterminate if,

$$Af_2 f_3 + B_2 f_3 + B_3 f_2 + C_1 = 0 \qquad (3.42)$$
$$\text{and} \quad B_1 f_2 f_3 + C_2 f_2 + C_3 f_3 + D = 0.$$

(Compare these relationships with those of Massau in equation (2.22), (2.23), and (2.24).)

If f_3 is eliminated from the equations (3.42) a quadratic equation in f_2 is obtained, namely

$$f_2^2(B_1 B_3 - AC_2) + f_2(B_1 C_1 + B_3 C_3 - B_2 C_2 - AD) + C_1 C_3 - B_2 D = 0.$$

Using d'Ocagne's notation this can be written,

$$E_2 f_2^2 - F_2 f_2 + G_2 = 0. \qquad (3.43)$$

A similar treatment which eliminates f_2 produces the quadratic,

$$E_3 f_3^2 - F_3 f_3 + G_3 = 0. \qquad (3.44)$$

It is important to note that the discriminant Δ of the equation

$$E_i f_i^2 - F_i f_i + G_i = 0$$

of which (3.43) and (3.44) are examples, is given by

$$\Delta = F_i^2 - 4E_i G_i$$

which is the same as the value given for Δ in d'Ocagne's original theory although the equation in that case was

$$E_i s_i^2 + F_i s_i + G_i = 0.$$

Returning to the theory of critical points it will be seen that in the equation of the third nomographic order,

$$A f_1 f_2 f_3 + B_1 f_2 f_3 + B_2 f_1 f_3 + B_3 f_1 f_2 + C_1 f_1 + C_2 f_2 + C_3 f_3 + D = 0$$

f_1 becomes indeterminate if f_2 and f_3 satisfy

$$E_i f_i^2 - F_i f_i + G_i = 0.$$

The roots of these equations denoted by f_2', f_2'' and f_3', f_3'' are given by,

$$2E_i f_i' - F_i = \sqrt{\Delta} \quad \text{and} \quad 2E_i f_i'' - F_i = -\sqrt{\Delta}$$

for $i = 2$ and 3.

Let the values of the variables corresponding to f_2', f_2'', f_3', and f_3'' be α_2', α_2'', α_3', and α_3''. Two further values, α_1' and α_1'', corresponding to f_1' and f_1'' can be found by applying conditions which will make f_2 or f_3 indeterminate. These values of course depend on Δ being greater than zero.

In Figure 3.30, the triangle $P_1 P_2 P_3$ is in the plane of the nomogram and is such that the variables α_i are

$$(\alpha_2', \alpha_3') \quad \text{at} \quad P_1,$$
$$(\alpha_3', \alpha_1'') \quad \text{at} \quad P_2,$$
$$\text{and} \quad (\alpha_1', \alpha_2'') \quad \text{at} \quad P_3.$$

The line $P_1 P_2$ is part of the line d_3 on which the values of α_3 are distributed. Similarly, $P_1 P_3$ is part of the line d_2 and $P_2 P_3$ is part of the line d_1. The indeterminacy is readily seen geometrically. For example, an alignment of α_2' and α_1'' lies along d_3 rendering α_3 indeterminate or, for a different type of indeterminacy, an alignment of α_2' and α_3'' passes through P_1 in an infinity of directions making α_1 indeterminate. There is an alternative arrangement, namely, $P_1(\alpha_2'', \alpha_3')$, $P_2(\alpha_3'', \alpha_1')$, and $P_3(\alpha_1'', \alpha_2')$ which is not homographically reducible to the first arrangement.

The purpose behind d'Ocagne's paper was to give an alternative, based on critical points, to the proof given in his *Traité* for the conditions under which (3.40) can be represented by a nomogram having three rectilinear supports and, when such representation is possible, to find the scales. It will be recalled that the condition was that $\Delta \geq 0$.

In the case where $\Delta > 0$ the scales are non-concurrent. This is the case arrived at above through a consideration of critical points. To find the scales in this case it is necessary to take three aligned points A_1, A_2, and A_3 on d_1, d_2, and d_3 respectively, so that they correspond to α_1, α_2, and α_3 (Figure 3.30). Each support will then have three dimension points, for example on d_1 there is P_3, A_1 and P_2. The scales can now be constructed by the usual method of projection.

If $\Delta = 0$ then
$$\alpha_1' = \alpha_1'' = \alpha_2' = \alpha_2'' = \alpha_3' = \alpha_3''$$
in which case P_1, P_2, and P_3 are all at the same point P in agreement with the earlier theory in which three rectilinear coincident scales are indicated by $\Delta = 0$.

If $\Delta < 0$, then the points P_1, P_2, and P_3 are imaginary and there is no rectilinear representation.

In his paper d'Ocagne begins with the assumption that a three rectilinear representation is possible; i.e., that $\Delta \geq$, and does not explain how the scales for $\Delta = 0$ are to be constructed. It is therefore not as complete an exposition as his earlier work.

Three months later d'Ocagne was able to show that, through a consideration of critical points, equation (3.40) could be transformed into Clark's irreducible form and therefore could be represented by a nomogram having two scales on

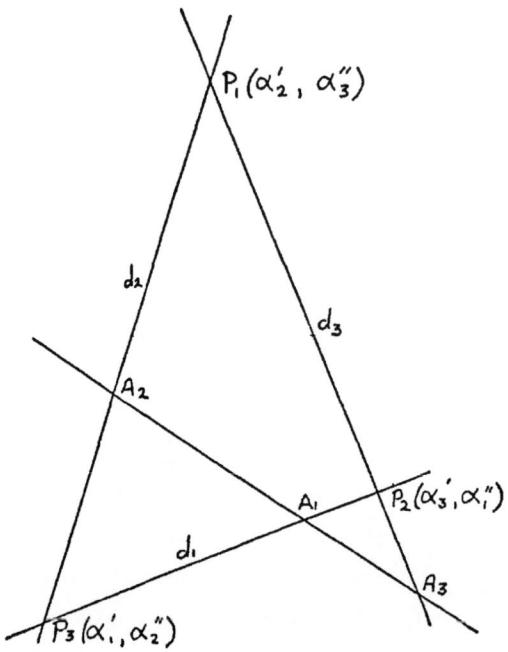

FIGURE 3.30

the same conic [**100**]. To demonstrate this, d'Ocagne starts from the equation

$$E_i f_i^2 - F_i f_i + G_i = 0,$$

the roots of which are the critical values of f_i. As has already been noted, the roots occur in two groups, f_i' and f_i'', and the three critical values of the same group give to

$$\phi_i = 2E_i f_i - F_i$$

the value $+\sqrt{\Delta}$ for f_i' or $-\sqrt{\Delta}$ for f_i''.

Putting

$$f_1 = \frac{\phi_1 + F_1}{2E_1} \quad \text{and} \quad f_2 = \frac{\phi_2 + F_2}{2E_2}$$

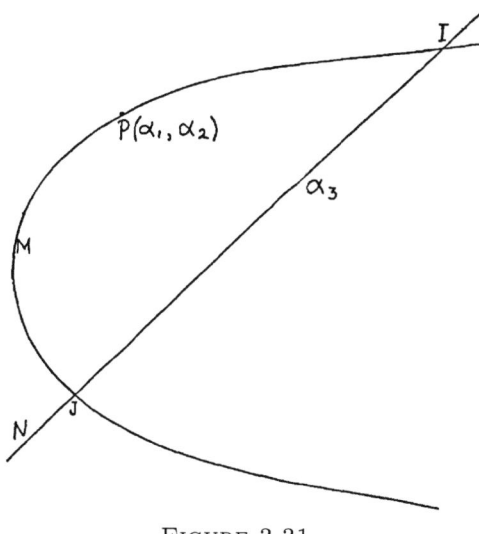

FIGURE 3.31

into equation (3.40) produces eventually an equation of the form,

$$\phi_1\phi_2 A_3 + (\phi_1 + \phi_2)C_3 + D_3 = 0$$

in which ϕ_1 and ϕ_2 are as defined above and A_3, C_3, and D_3 are linear functions of f_3. Furthermore, any point on the conic must represent a pair of critical values, one value for each of the variables that it carries, since this pair must render the third variable indeterminate, as an examination of Figure 3.31 will show. In this sketch it is assumed that α_1 and α_2 are on the conical support M and α_3 on the straight line N. It follows that the values of α_1 and α_2 corresponding to the same point of the conical support are linked by,

$$2E_1 f_1 - F_1 = 2E_2 f_2 - F_2.$$

If the support N cuts the support M in I and J; i.e., when $\Delta > 0$, then the three values of the variables at both I and J must be critical. For example, I may correspond to the critical values f_1', f_2', and f_3', and J to f_1'', f_2'', and f_3''. The effect of this is that the support N of α_3 can be constructed from a knowledge of the critical values of either α_1 or α_2. One normal alignment of α_1, α_2, and α_3 will give a third point on N and hence the scale of α_3 can be constructed.

In the case where $\Delta = 0$, I and J coincide and N is a tangent to M. If $\Delta < 0$ the critical values are imaginary and N does not cut M in any real points.

There is little practical value in this aspect of d'Ocagne's work since it is only applied to third order equations and therefore lacks the more general appeal of Clark's work. However, it does provide the intellectual satisfaction gained from drawing together two separately developed pieces of work.

As an illustration consider,

$$f_1 f_2 + f_1 f_3 - f_2 f_3 - 2f_1 + 2f_2 - f_3 = 0$$

which in Clark's determinant form is,

$$\begin{vmatrix} f_1 & f_1^2 & 1 \\ -f_2 & f_2^2 & 1 \\ f_3 - 2 & f_3 & 1 \end{vmatrix} = 0.$$

A similar example was given in the section on Clark's work. In that example it was seen that the scales of f_1 and f_2 were in opposite directions from the axis of the parabola $y = x^2$ (Figure 3.11). In terms of d'Ocagne's notation the following apply:

$$A = 0, \; B_1 = -1, \; B_2 = B_3 = 1, \; C_1 = -2, \; C_2 = 2, \; C_3 = -1, \text{ and } D = 0,$$

giving $F_0 = 3$ and

$$\begin{aligned} E_1 = -1, && F_1 = -1, && G_1 = 2, && \Delta = 9; \\ E_2 = 1, && F_2 = -1, && G_2 = -2, && \Delta = 9; \\ E_3 = 1, && F_3 = 5, && G_3 = 4, && \Delta = 9. \end{aligned}$$

The equations giving critical values are,

$$\begin{aligned} f_1^2 - f_1 - 2 = 0 & \quad \text{giving} \quad & f_1' &= 2 \text{ and } f_1'' = -1, \\ f_2^2 + f_2 - 2 = 0 & \quad \text{giving} \quad & f_2' &= 1 \text{ and } f_2'' = -2, \\ \text{and} \quad f_3^2 - 5f_3 + 4 = 0 & \quad \text{giving} \quad & f_3' &= 4 \text{ and } f_3'' = 1 \end{aligned}$$

and for any point on the parabola

$$2E_1 f_1 - F_1 = 2E_2 f_2 - F_2$$

leads to $f_1 = -f_2$ which is as expected. It is also seen that the set of values for f'; i.e., 2, 1, and 4, satisfy the original relationship and therefore correspond to the point I (or J) and similarly for the values of f''; i.e., -1, -2, and 1.

A similar analysis can be carried out in the case in which $A = 1$, $B_1 = B_2 = B_3$ and $C_1 = C_2 = C_3$, which is the case shown by Clark to lead to cubic nomograms. In this case the equations leading to the critical values are the same for all functions, the critical values coinciding at the double point. If $\Delta > 0$ the curve is crunodal, if $\Delta = 0$ it is cuspidal and if $\Delta < 0$ it is acnodal.

These points can be illustrated with reference to three examples given in the section on Clark's work.

3.5.1. Crunodal Form. $f_1 f_2 f_3 - 1 = 0$ which produced the folium of Descartes, of crunodal form, Figure 3.27.

$$A = 1, B_i = 0, C_i = 0, \text{ and } D = -1$$

giving

$$F_0 = 1, E_i = 0, F_i = 1, \text{ and } G_i = 0$$

from which $\Delta = 1$ and the equation for critical values is $f_i = 0$.

3.5.2. Cuspidal Form. $f_1 f_2 + f_2 f_3 + f_1 f_3 = 0$ which led to the curve $x^3 = y^2$ of cuspidal form, Figure 3.28.

$$A = 0, B_i = 1, C_i = 0, \text{ and } D = 0$$

giving

$$F_0 = 0, E_i = -1, F_i = 0, \text{ and } G_i = 0$$

from which $\Delta = 0$ and the equation for critical values is $f_i^2 = 0$.

3.5.3. Acnodal Form. $f_1 f_2 f_3 - (f_1 + f_2 + f_3) = 0$ which led to the curve $y = x(x^2 + 1)^{-1}$ of acnodal form, Figure 3.29.

$$A = 1, B_i = 0, C_i = -1, \text{ and } D = 0$$

giving

$$F_0 = 0, E_i = -1, F_i = 0, \text{ and } G_i = -1$$

from which $\Delta = -4$ and the equation for critical values is $f_i^2 = -1$.

Also in 1907, d'Ocagne used his notion of critical points to reproduce a result of Clark in which it is required to find the condition under which

$$f_1 f_2 A_3 + f_1 B_3 + f_2 C_3 + D_3 = 0$$

is representable by a nomogram of genus 1 [**101**]. Here A_3, B_3, C_3, and D_3 are given by

$$A_3 = a_0 f_3 + b_0 \phi_3 + c_0 \psi_3,$$
$$B_3 = a_1 f_3 + b_1 \phi_3 + c_1 \psi_3,$$
$$C_3 = a_2 f_3 + b_2 \phi_3 + c_2 \psi_3,$$
$$\text{and} \quad D_3 = a_3 f_3 + b_3 \phi_3 + c_3 \psi_3.$$

Clark's result is,

$$\begin{vmatrix} a_1 & a_2 & a_3 \\ b_1 & b_2 & b_3 \\ c_1 & c_2 & c_3 \end{vmatrix} \times \begin{vmatrix} a_0 & a_1 & a_2 \\ b_0 & b_1 & b_2 \\ c_0 & c_1 & c_2 \end{vmatrix} - \begin{vmatrix} a_0 & a_2 & a_3 \\ b_0 & b_2 & b_3 \\ c_0 & c_2 & c_3 \end{vmatrix} \times \begin{vmatrix} a_0 & a_1 & a_3 \\ b_0 & b_1 & b_3 \\ c_0 & c_1 & c_3 \end{vmatrix} = 0.$$

D'Ocagne gets the same result by firstly writing the equation in the form,

$$f_3(a_0 f_1 f_2 + a_1 f_1 + a_2 f_2 + a_3) +$$
$$\phi_3(b_0 f_1 f_2 + b_1 f_1 + b_2 f_2 + b_3) +$$
$$\psi_3(c_0 f_1 f_2 + c_1 f_1 + c_2 f_2 + c_3) = 0.$$

He then points out that the two rectilinear scales intersect when the values of f_1 and f_2 corresponding to this point are critical, rendering the value of the third variable indeterminate. Thus the coefficients of f_3, ϕ_3, and ψ_3 must all be zero. From this he quotes a condition, based on a procedure given in his *Traité*, which yields Clark's result.

Finally, one must note the use of the concept of critical points to express an equation of the form,

$$F_1 F_{23} + G_1 G_{23} + H_1 H_{23} = 0,$$

or nomographic order 2 at the most with respect to the variable α_1, in the form,

$$\begin{vmatrix} F_1 & G_1 & H_1 \\ F_2 & G_2 & H_2 \\ F_3 & G_3 & H_3 \end{vmatrix} = 0.$$

This is due to Farid Boulad who has already been noted in connection with alignment nomograms used by Egyptian Railways to determine the strengths of railway bridges [10]. His method is complicated in detail but in essence is similar to those given above.

3.6. The Period of Consolidation

With the exception of many attempts to solve the important theoretical problem of nomography, the development of the subject now adopted a much more leisurely pace. In fact for some forty years until the late 1950's the publications on the subject were just those that one associates with a developed discipline for they mainly fall into the following categories:

a. treatises and text books
b. papers relating to the early history of the subject
c. papers on theoretical aspects

Amongst the treatises and textbooks must be noted the new edition of d'Ocagne's *Traité* which appeared in 1921 [103]; the two volume work of Soreau which was also published in 1921 [127]; the publication of d'Ocagne's lectures on pure and applied geometry given at the École Polytechnique [102], and, in English, the first edition of *The Nomogram* by Allcock and Jones [1].

Those papers relating to the history of the subject tended to be rather superficial. There were not many of them and, although useful in parts, cannot be said to have contributed greatly to the record of the development of the subject. Many are biased towards the author's own work and the major contributions of Massau and Clark are lightly dealt with. Amongst these papers it is worth noting those of d'Ocagne [98] and [104], and of Lallamand [71]. Of interest to future researchers in nomography is a note by d'Ocagne describing the nomographic archives at the École des Ponts et Chaussées [105]. The papers on theoretical aspects are the subject of the next chapter.

Two short papers by Wladimir Margoulis which appeared in 1922 and 1923 are worthy of mention [80] and [81]. The first would appear to be a link between Lallemand's hexagonal nomogram and the work of the Russian G.S. Khovanskii which appeared in 1959 [60]. It is on the construction of nomograms using oriented transparencies and describes canonical forms of equations representable by this method. The second paper deals with the general theory of the representation of equations using moveable elements. Both papers show that some development of technique was still taking place during this period of consolidation but it is not necessary for the development of this thesis to discuss those techniques in detail.

CHAPTER 4

Anamorphosis and Alignment Nomograms

4.1. The Problem

It has already been established that the theoretical problems of anamorphosis and of the construction of alignment nomograms can be stated as the problem of expressing the relationship,

$$F(x, y, z) = 0 \tag{4.1}$$

in the form,

$$\begin{vmatrix} f_1(x) & g_1(x) & 1 \\ f_2(y) & g_2(y) & 1 \\ f_3(z) & g_3(z) & 1 \end{vmatrix} = 0. \tag{4.2}$$

The attempt of Duporcq to solve this problem has already been considered as have the works of Saint-Robert, Massau and Lecornu which deal with a less general form of the problem. The present chapter looks at later and more substantial attempts to solve this problem.

These later attempts began with one advantage for earlier investigators had pointed out two possible lines of attack. The first approach was that which had been followed by Saint-Robert, Massau and Lecornu and indicated by d'Ocagne in his *Nomographie*. It was to observe that, as d'Ocagne said,

> "The common character of all equations susceptible of reverting to the determinant form is that they express themselves by partial differential equations obtained as a result of the elimination of the arbitrary functions which enter into that form."

171

The second approach had been suggested by Duporcq and rested on the fact that the form of the expanded determinant must be,

$$P_1(x)R_1(y, z) + P_2(x)R_2(y, z) + P_3(x)R_3(y, z)$$

at its most general, if expanded along the x row, with parallel expressions if expanded along the y row or the z row.

4.2. Grönwall's Approach

Grönwall approached the problem through partial differential equations. His paper appeared in a French journal although he himself is described as being from Chicago [**51**]. The forty three pages of intricate mathematics, leaning heavily on the theory of partial differential equations, serve to indicate the complexity of the problem being investigated and it is only possible here to give a general view of the work, highlighting a few points that seem relevant to this thesis. At some points the paper shows clearly its relationship to the work of Saint-Robert and Massau.

The main result of the paper gives a necessary and sufficient condition for equation (4.1) to be reduced to the determinant form (4.2). It is that the following two partial differential equations should have a common solution C.

$$M\frac{\partial^2 C}{\partial x^2} + 2\frac{\partial^2 C}{\partial y^2} = \left(MC - 2\frac{\partial M}{\partial x}\right)\frac{\partial C}{\partial x} \qquad (4.3)$$
$$+ 2C\frac{\partial C}{\partial y} + \frac{\partial^2 C}{\partial x^2} + \left(\frac{\partial N}{\partial x} - \frac{\partial^2 M}{\partial x^2}\right)C - \frac{\partial^2 N}{\partial x^2}$$

$$2M\frac{\partial^2 C}{\partial x \partial y} + \frac{\partial^2 C}{\partial x^2} = 2\left(M^2 C + MN - \frac{\partial M}{\partial y}\right)\frac{\partial C}{\partial x} \qquad (4.4)$$
$$+ \left(MC + N - 2\frac{\partial M}{\partial x}\right)$$
$$+ 2M\frac{\partial^2 C}{\partial s^2} + 2\left(N\frac{\partial M}{\partial x} + M\frac{\partial N}{\partial x} - \frac{\partial^2 M}{\partial x \partial y}\right)C$$
$$+ 2N\frac{\partial N}{\partial x} - 2\frac{\partial^2 N}{\partial x \partial y}$$

(In the original paper the second equation contains two typographical errors). In these equations, M and N are given by,

$$M = -\frac{\partial z}{\partial y}\bigg/\frac{\partial z}{\partial x} \qquad (4.5)$$

and

$$N = \frac{\partial M}{\partial x} + M^{-1}\frac{\partial M}{\partial y} \qquad (4.6)$$

The quantity C is important to much of Grönwall's theory and is the subject of his second result. This is that all equations (4.2) which belong to the same value of C can be obtained, one from another, by a homographic projection and that inversely two homographic equations lead to the same value of C.

The complexity of equations (4.3) and (4.4) is obvious. However, in some cases it is not necessary to solve these equations in order to find C, for the theory develops along lines which relate particular types of nomograms to particular relationships between the derivatives of M, N, and C and a quantity D, (defined below), which enable C to be found. The process by which C is found ensures that it satisfies equations (4.3) and (4.4). D is given

$$D = MC + N. \qquad (4.7)$$

It will be noted that M bears a similarity to the quantity R defined by Saint-Robert (see equation (2.10)). In fact

$$R = -\frac{1}{M}.$$

Saint-Robert was able to give a condition for $F(x, y, z) = 0$ to be reducible to the form $Z(z) = X(x) + Y(y)$. It was

$$\frac{\partial^2 \ln R}{\partial x \partial y} = 0,$$

the condition known as Saint-Robert's criterion.

Grönwall gives a necessary and sufficient condition for equation (4.1) to be represented by a nomogram having three rectilinear scales. It is

$$\frac{\partial^2 \ln M}{\partial x \partial y} = 0. \qquad (4.8)$$

The parallel is obvious.

Two other results of Grönwall are that,

(i) The necessary and sufficient condition for the x scale to be rectilinear is that

$$\frac{\partial C}{\partial y} + 2\frac{\partial D}{\partial x} = 0 \tag{4.9}$$

taken with (4.7) and the pair,

$$2\frac{\partial^2 C}{\partial x \partial y} + \frac{\partial^2 D}{\partial x^2} - C\left(2\frac{\partial C}{\partial y} + \frac{\partial D}{\partial x}\right) = 0 \tag{4.10}$$

$$\text{and} \quad \frac{\partial^2 C}{\partial y^2} + 2\frac{\partial^2 D}{\partial x \partial y} - D\left(\frac{\partial C}{\partial y} + 2\frac{\partial D}{\partial x}\right) = 0.$$

(ii) The necessary and sufficient condition for the y scale to be rectilinear is,

$$2\frac{\partial C}{\partial y} + \frac{\partial D}{\partial x} = 0 \tag{4.11}$$

taken with (4.7) and (4.10).

Combining the above results gives the necessary and sufficient condition for the scales of x and y both to be rectilinear. It is,

$$\frac{\partial C}{\partial y} = \frac{\partial D}{\partial x} = 0 \tag{4.12}$$

taken with (4.7).

When C is known, Grönwall has a procedure for finding the functions f_i and g_i. It is not a simple procedure involving, as it does, the need to find a fundamental solution set of the following system of partial differential equations:

$$\frac{\partial^2 w}{\partial x^2} = \frac{C}{3}\frac{\partial w}{\partial x} + \frac{1}{3}\left(\frac{2}{3}C^2 - \frac{\partial C}{\partial x}\right)w,$$

$$\frac{\partial^2 w}{\partial x \partial y} = \frac{-D}{3} - \frac{C}{3}\frac{\partial w}{\partial y} + \frac{1}{3}\left(\frac{-CD}{3} + \frac{\partial C}{\partial y} + \frac{\partial D}{\partial x}\right)w,$$

$$\text{and} \quad \frac{\partial^2 w}{\partial y^2} = \frac{D}{3}\frac{\partial w}{\partial y} + \frac{1}{3}\left(\frac{2}{3}D^2 - \frac{\partial D}{\partial y}\right)w.$$

The method may be of use as a last resort but I am doubtful of its practical value in the majority of cases.

However, some of ideas work well, particularly in simple cases. Consider

$$F(x, y, z) = z - x^2 y^2 = 0$$

which was used to illustrate Massau's approach. Grönwall first notes that the Jacobian,

$$\frac{\partial(z, \phi(x) + \psi(y))}{\partial(x, y)} = 0$$

implies a relationship

$$\chi(z) + \phi(x) + \psi(y) = 0,$$

which can be written as,

$$\begin{vmatrix} \phi(x) & -1 & 1 \\ \psi(y) & 1 & 1 \\ -\frac{1}{2}\chi(z) & 0 & 1 \end{vmatrix} = 0.$$

The procedure is to obtain M from $z = x^2 y^2$; i.e.,

$$M = -\frac{\partial z}{\partial y} \bigg/ \frac{\partial z}{\partial x} = -\frac{2x^2 y}{2xy^2} = -\frac{x}{y}.$$

(Note that $\frac{\partial^2 \ln M}{\partial x \partial y} = 0$.)

M is then expressed as $M = \alpha(x)\,\beta(y)$ and two arbitrary functions are obtained as follows,

$$\phi(x) = \int \frac{dx}{\alpha(x)} = -\ln x \quad \text{and} \quad \psi(y) = -\int \beta(y)\,dy = -\ln y$$

hence

$$M = -\frac{x}{y} = -\frac{\psi'(y)}{\phi'(x)} = -\frac{\partial z}{\partial y} \bigg/ \frac{\partial z}{\partial x},$$

the last two terms giving the Jacobian above. Therefore,

$$\chi(z) = \ln x + \ln y = \ln(xy) = \ln \sqrt{z} = \frac{1}{2} \ln z$$

and

$$F(x, y, z) = \begin{vmatrix} -\ln x & -1 & 1 \\ -\ln y & 1 & 1 \\ -\frac{1}{4}\ln z & 0 & 1 \end{vmatrix} = 0.$$

In considering the case of two rectilinear scales with the third scale some other curve, Grönwall is following in the footsteps of Massau and Lecornu. The form examined by them was,

$$Z_1(z)X(x) + Z_2(z)Y(y) = 1$$

which can be written

$$\begin{vmatrix} X(x)^{-1} & 0 & 1 \\ 0 & Y(y)^{-1} & 1 \\ Z_1(z) & Z_2(z) & 1 \end{vmatrix} = 0$$

showing clearly that the scales of x and y are rectilinear while that of z is a curve.

In this case, Grönwall's analysis gives,

$$C = -\frac{\partial}{\partial y}\left(\frac{1}{M}\frac{\partial N}{\partial x}\right) \bigg/ \frac{\partial^2 \ln M}{\partial x \partial y} \ . \tag{4.13}$$

The necessary and sufficient conditions for the scales of x and y to be rectilinear while that of z is a curve are that C should satisfy the equations,

$$\frac{\partial C}{\partial y} = \frac{\partial (MC + N)}{\partial x} = 0. \tag{4.14}$$

Grönwall states that condition (4.14) had been obtained in a totally different manner by Massau and this is so. I have compared both developments and find that Massau'a conditions (2.47)

$$\frac{\partial \lambda}{\partial x} = 0 \quad \text{and} \quad \frac{\partial \mu}{\partial y} = 0$$

and Grönwall's conditions

$$\frac{\partial (MC + N)}{\partial x} = 0 \quad \text{and} \quad \frac{\partial C}{\partial y} = 0$$

are identical.

At the end of his paper Grönwall turns his attention to Clark's conical nomograms. He finds that, for $F(x, y, z) = 0$ to be reduced to the form

$$\begin{vmatrix} f_1(x) & f_1^2(x) & 1 \\ f_2(y) & f_2^2(y) & 1 \\ f_3(z) & g_3(z) & 1 \end{vmatrix} = 0. \tag{4.15}$$

C must be given by

$$C = -\frac{\partial}{\partial y}\left(\frac{1}{M}\frac{\partial N}{\partial x}\right) \Big/ \frac{\partial^2 \ln M}{\partial x \partial y} \tag{4.16}$$

which is the same expression as (4.13).

The necessary and sufficient condition for (4.1) to be reduced to (4.15) is that C, given by (4.16), must satisfy,

$$\frac{\partial C}{\partial y} = \frac{\partial(MC + N)}{\partial x} = \frac{\partial D}{\partial x} \neq 0 \tag{4.17}$$

$$\text{and} \quad \frac{\partial^2 C}{\partial x \partial y} = C\frac{\partial C}{\partial y}.$$

It is worth observing that some relationships for which conical charts can be constructed, such as $z = xy$, would make the denominator of (4.16) zero. This can be seen, for example, in (3.33) with $B_3 = 0$, $C_3 = -f_3$, and $A_3 = 1$.

In the introduction to his paper Grönwall states that in a subsequent work he intends to make explicit the common integral of the partial differential equations (4.3) and (4.4). I have not succeeded in finding out whether he ever did this.

4.3. Kellogg's Approach

Although Kellogg's paper was published three years after Grönwall's, it can be regarded as a response to Grönwall if one takes note of the following facts. Grönwall, from Chicago, published his paper in France, in French, in 1912. Kellogg's paper was published in Germany, in English, in 1915, but the paper is dated February 23, 1913 and was written in Columbia, Missouri, [59]. The delay in publication was no doubt aggravated by the tension and subsequent war in Europe. It therefore seems likely that Kellogg wrote his paper within months of reading Grönwall's paper to which he refers in his introduction. Also in his introduction is this statement,

> "If I venture a contribution to the subject, it is because the criteria which I have found seem to leave little to be desired in point of simplicity of application, involving as they do merely differentiations and the determination of the ranks of matrices".

In order that this statement should not mislead, one must bear in mind the known difficulty of the problem.

Kellogg's approach is quite different from previous ones. The underlying concept is that of linear dependence and in particular the linear dependence of functions of several variables. The relevance of this concept can be seen if one considers what he calls the irreducible case of the nomographic problem by which he means the case in which $F(x, y, z)$ can be expressed in the form,

$$P_1(x)R_1(y, z) + P_2(x)R_2(y, z) + P_3(x)R_3(y, z)$$

but not in any reduced form such as

$$P_4(x)R_4(y, z) + P_5(x)R_5(y, z).$$

This irreducible form requires that $P_1(x)$, $P_2(x)$, and $P_3(x)$ should be linearly independent and also that $R_1(y, z)$, $R_2(y, z)$, and $R_3(y, z)$ should be linearly independent. While the condition for the linear independence of the P's is well enough known, that for the R's is not often encountered. Kellogg describes the latter as "a result of some interest which I have thus far failed to meet with in print." The condition relates the number of linear independent functions to the rank of a matrix which has for its elements the functions and certain of their partial derivatives.

Starting with the irreducible form,

$$F(x, y, z) = P_1(x)R_1(y, z) + P_2(x)R_2(y, z) + P_3(x)R_3(y, z) \qquad (4.18)$$

Kellogg notes that F must satisfy an ordinary homogeneous third order differential equation in x having coefficients which depend only on x. This fact can easily be confirmed by differentiating (4.18) three times partially with respect to x and stating the condition for the consistency of the resulting equations. This gives,

$$\begin{vmatrix} F & P_1 & P_2 & P_3 \\ F_x & P_1' & P_2' & P_3' \\ F_{xx} & P_1'' & P_2'' & P_3'' \\ F_{xxx} & P_1''' & P_2''' & P_3''' \end{vmatrix} = 0.$$

This differential equation can be regarded as a homogeneous linear relationship between F, F_x, F_{xx}, and F_{xxx}, with coefficients that do not contain y and z. To this linear relationship Kellogg applies his condition for linear independence. His result is that the necessary and sufficient condition for $F(x, y, z)$ to be expressed in the irreducible form (4.18) is that a 4 by 10 matrix N should be of rank less than four and that a matrix N' obtained from N by deleting the last row and the last four columns, should be of rank three. N and N' are given in Appendix B.

Next it is necessary to find the functions in (4.18). The P_i's are obtained by forming the differential equation satisfied by F and finding three independent solutions to it. These independent solutions are $P_1(x)$, $P_2(x)$, and $P_3(x)$. Once the P_i's have been found the R_i's can often be found by inspection. If they cannot be so found, then it is necessary to differentiate (4.18) twice with respect to x and solve the linear system formed by the two resulting equations and (4.18).

The method so far is illustrated in the following example:

$$F(x, y, z) = e^x y^2 - e^x z^3 - xe^y + xe^z + x^2 z^3 e^y - e^z x^2 y^2.$$

The matrices N and N' are both of rank three, although demonstrating this is most tedious. By contrast, forming the differential equation which is satisfied by F is quite simple. It is,

$$(x^2 - 2x + 2)F_{xxx} - x^2 F_{xx} + 2xF_x - 2F = 0.$$

In this case three independent solutions are easy to spot, they are e^x, x, and x^2. Rearranging the equation, the R_i's immediately reveal themselves. We have

$$P_1(x) = e^x, \qquad R_1(y, z) = y^2 - z^3,$$
$$P_2(x) = x, \qquad R_1(y, z) = e^z - e^y,$$
$$\text{and} \quad P_3(x) = x^2, \qquad R_1(y, z) = e^y z^3 - e^z y^2.$$

Having established that the form (4.18) is possible it is next necessary to investigate whether a determinant form is possible; i.e., is it possible to write,

$$F(x, y, z) = \begin{vmatrix} f_1(x) & g_1(x) & h_1(x) \\ f_2(y) & g_2(y) & h_2(y) \\ f_3(z) & g_3(z) & h_3(z) \end{vmatrix} ? \tag{4.19}$$

Since $f_1(x)$, $g_1(x)$, and $h_1(x)$ must be solutions to the differential equation which has $P_1(x)$, $P_2(x)$, and $P_3(x)$ as independent solutions, a homographic transformation can be found which will make the first row $P_1(x)$, $P_2(x)$, and $P_3(x)$. Therefore, the form we have is,

$$F(x, y, z) = \begin{vmatrix} P_1(x) & P_2(x) & P_3(x) \\ f_2(y) & g_2(y) & h_2(y) \\ f_3(z) & g_3(z) & h_3(z) \end{vmatrix} \tag{4.20}$$

that is

$$R_1 = \begin{vmatrix} g_2 & h_2 \\ g_3 & h_3 \end{vmatrix}, \quad R_2 = \begin{vmatrix} h_2 & f_2 \\ h_3 & f_3 \end{vmatrix}, \quad \text{and} \quad R_3 = \begin{vmatrix} f_2 & g_2 \\ f_3 & g_3 \end{vmatrix}.$$

There exists a homogeneous linear relation between R_1, R_2 and R_3 with coefficients depending only on y, for example,

$$f_2 R_1 + g_2 R_2 + h_2 R_3 = 0.$$

A similar relationship exists having coefficients which depend only on z.

The necessary and sufficient conditions for these relationships are

$$\begin{vmatrix} R_1 & R_2 & R_3 \\ R_{1z} & R_{2z} & R_{3z} \\ R_{1zz} & R_{2zz} & R_{3zz} \end{vmatrix} = 0 \quad \text{and} \quad \begin{vmatrix} R_1 & R_2 & R_3 \\ R_{1y} & R_{2y} & R_{3y} \\ R_{1yy} & R_{2yy} & R_{3yy} \end{vmatrix} = 0. \quad (4.21)$$

Thus the coefficients are proportional to the minors of those Wronskians. Suppose them to be found and to be Y_1, Y_2, and Y_3 and Z_1, Z_2, and Z_3 respectively. Then,

$$Y_1 R_1 + Y_2 R_2 + Y_3 R_3 = 0$$

$$\text{and} \quad Z_1 R_1 + Z_2 R_2 + Z_3 R_3 = 0.$$

R_1, R_2, and R_3 can now be expressed as

$$R_1 = \rho \begin{vmatrix} Y_2 & Y_3 \\ Z_2 & Z_3 \end{vmatrix}, \quad R_2 = \rho \begin{vmatrix} Y_3 & Y_1 \\ Z_3 & Z_1 \end{vmatrix}, \quad \text{and} \quad R_3 = \rho \begin{vmatrix} Y_1 & Y_2 \\ Z_1 & Z_2 \end{vmatrix}$$

where ρ is independent of x.

In order that the determinant form (4.20) is possible it is necessary for ρ to be a function of y times a function of z. That this must be so is easily seen by assuming that $\rho = \sigma(y)\tau(z)$. Then,

$$R_1 = \sigma(y)\tau(z)\left(Y_2(y)Z_3(z) - Y_3(y)Z_2(z)\right)$$

$$= g_2(y)h_3(z) - h_2(y)g_3(z).$$

Therefore, if $\rho = \sigma(y)\tau(z)$, it follows that,

$$\frac{\partial^2 \ln \rho}{\partial y \partial z} = 0. \quad (4.22)$$

Kellogg's conclusion is that, having satisfied the conditions for the irreducible form (4.18), the necessary and sufficient conditions for $F(x, y, z)$ to be expressed in the determinant form (4.20) are those expressed by (4.21) and (4.22).

In the example given above, the Wronskians of (4.21) become

$$\begin{vmatrix} y^2 - z^3 & e^z - e^y & e^y z^3 - e^z y^2 \\ -3z^2 & e^z & 3z^2 e^y - e^z y^2 \\ -6z & e^z & 6z e^y - e^z y^2 \end{vmatrix}$$

and

$$
\begin{vmatrix}
y^2 - z^3 & e^z - e^y & e^y z^3 - e^z y^2 \\
2y & -e^y & e^y z^3 - 2y e^z \\
2 & -e^y & e^y z^3 - 2e^z
\end{vmatrix}
$$

both of which vanish; and in this case

$$
\rho = 6e^y (1 - y) z e^z (2 - z)
$$

giving

$$
\frac{\partial^2 \ln \rho}{\partial y \partial z} = 0.
$$

Kellogg is more concerned with the criteria for expressing a function in the form (4.20) than with actually finding that form and he does not spell out how the components of R_1, R_2, and R_3 are to be found. However, once it is known that the form is possible these components can be found by inspection. In the example the final form is

$$
\begin{vmatrix}
e^x & x & x^2 \\
e^y & y^2 & 1 \\
e^z & z^3 & 1
\end{vmatrix}
$$

Kellogg deals with the two simpler cases,

$$
F(x, y, z) = P_4(x) R_4(y, z) + P_5(x) R_5(y, z)
$$

and

$$
F(x, y, z) = P_6(x) R_6(y, z)
$$

in a similar manner.

Finally, in order to remove the need to solve differential equations, Kellogg expresses his criteria in terms of $F(x, y, z)$ and its derivatives. These alternative forms are of little practical value, in most cases, because of the complexity of the expressions.

The merit of Kellogg's approach is that it attacks the problem in a new way. His criteria are rather complicated for any case where the existence of the determinant form is in doubt, that is in any case complex enough to require the criteria to be tested. However, if the form is known to exist, the method for finding the components is a reasonable enough procedure as the earlier example illustrates.

4.4. Warmus and Nomographic Functions

An exhaustive attack on the problem of anamorphosis came from Poland in 1959 [**134**]. The author was Mieczyslaw Warmus who acknowledged the earlier works of Duporcq, Grönwall and Kellogg, making special reference to the latter's existence criteria, considering them to be "unnecessarily complicated" and leading to "computations too long and troublesome for practical use".

Warmus's approach is algebraic. Rather than seek for conditions under which $F(x, y, z)$ can be expressed in a nomographic form, he attempts to classify those forms which are suitable and gives an elaborate computation scheme which either leads to a determinant form or indicates that such a form is not possible.

There are two important preliminary ideas on which he erects his work. Firstly, that of linear independence of functions of one variable. His theorem on this is of some interest since it depends on the existence of numbers, within a given range, which satisfy a certain condition. This concept of the existence of a set of numbers satisfying certain conditions threads its way through the whole of his work. The theorem states that the functions $T_i(t)$ are linearly independent if, and only if, there exist numbers t_i, within the given range, such that,

$$\begin{vmatrix} T_1(t_1) & T_2(t_1) & \ldots & T_n(t_1) \\ T_1(t_2) & T_2(t_2) & \ldots & T_n(t_2) \\ \ldots & & & \ldots \\ T_1(t_n) & T_2(t_n) & \ldots & T_n(t_n) \end{vmatrix} \neq 0.$$

The second idea is that of the rank of a function. This is initially defined in terms of a function of two variables as follows:

$G(u, v)$ is of rank $n > 1$ if, and only if, there exist functions of u, $U_1, U_2, \ldots, U_n$ and functions of v, $V_1, V_2, \ldots, V_n$ such that

$$G(u, v) = U_1 V_1 + U_2 V_2 + \ldots + U_n V_n$$

there being no functions $\bar{U}_1, \ldots, \bar{U}_{n-1}, \bar{V}_1, \ldots, \bar{V}_{n-1}$ such that

$$G(u, v) = \bar{U}_1 \bar{V}_1 + \bar{U}_2 \bar{V}_2 + \ldots + \bar{U}_{n-1} \bar{V}_{n-1}.$$

The notion of the rank of a function is developed and a procedure evolved for finding the rank of a given function and the functions $U_1, U_2, \ldots, U_n$ and $V_1, V_2, \ldots, V_n$ into which it can be decomposed. It is a feature of such a decomposition that the functions must be linearly independent as also must be the functions. The influence of Kellogg's work can be detected at this stage. A result which has a geometric parallel in a homographic transformation is obtained. Stated for $n = 3$ it is as follows:

If

$$U_1V_1 + U_2V_2 + U_3V_3 = \bar{U}_1\bar{V}_1 + \bar{U}_2\bar{V}_2 + \bar{U}_3\bar{V}_3$$

in which the U's are linearly independent and the V's are linearly independent, then a matrix A must exist, where,

$$A = \begin{pmatrix} a_{11} & a_{12} & a_{13} \\ a_{21} & a_{22} & a_{23} \\ a_{31} & a_{32} & a_{33} \end{pmatrix}$$

such that

$$\bar{U}_i = a_{i1}U_1 + a_{i2}U_2 + a_{i3}U_3$$

$$\text{and} \quad \bar{V}_i = b_{i1}V_1 + b_{i2}V_2 + b_{i3}V_3$$

where $a = \det A \neq 0$ and $b_{ij} = a_{ij}^*/a$, a_{ij}^* being the cofactor of a_{ij}.

Warmus extends the notion of rank to functions of three variables since the nomographic problem is concerned with the function $F(x, y, z)$. The extension takes the form of grouping two of the variables together, say (y, z), so that we have the definition,

$F(x, y, z)$ is said to be of rank n with respect to x if, and only if, when considered as a function of the two variables x and (y, z) it is of rank n. Similar definitions apply for rank with respect to y and with respect to z.

Warmus's notion of a nomographic function can now be defined. $F(x, y, z)$ is said to be nomographic if, and only if, the following apply:

Condition 1: There exist functions $X_i(x)$, $Y_i(y)$, and $Z_i(z)$ and x, y, and z in appropriate ranges such that

$$F(x, y, z) = \begin{vmatrix} X_1(x) & X_2(x) & X_3(x) \\ Y_1(y) & Y_2(y) & Y_3(y) \\ Z_1(z) & Z_2(z) & Z_3(z) \end{vmatrix}$$

Condition 2: $F(x, y, z)$ is of rank greater than 1 with respect to each of the variables x, y, and z.

He calls the determinant form in Condition 1 a *Massau determinant*. The purpose of Condition 2 is to exclude trivial cases.

Warmus is next led to consider the equivalence of two Massau forms of $F(x, y, z)$ and gives the following definition.

The two Massau forms of $F(x, y, z)$,

$$\begin{vmatrix} X_1 & X_2 & X_3 \\ Y_1 & Y_2 & Y_3 \\ Z_1 & Z_2 & Z_3 \end{vmatrix} \quad \text{and} \quad \begin{vmatrix} \bar{X}_1 & \bar{X}_2 & \bar{X}_3 \\ \bar{Y}_1 & \bar{Y}_2 & \bar{Y}_3 \\ \bar{Z}_1 & \bar{Z}_2 & \bar{Z}_3 \end{vmatrix}$$

are equivalent if, and only if, there exists a matrix of numbers

$$A = \begin{pmatrix} a_{11} & a_{12} & a_{13} \\ a_{21} & a_{22} & a_{23} \\ a_{31} & a_{32} & a_{33} \end{pmatrix}$$

with $a = \det A \neq 0$, and if there also exist two numbers d_1 and d_2 satisfying the condition $a\, d_1 d_2 = 1$ such that

$$\begin{pmatrix} X_1 & X_2 & X_3 \\ Y_1 & Y_2 & Y_3 \\ Z_1 & Z_2 & Z_3 \end{pmatrix} = \begin{pmatrix} \bar{X}_1 & \bar{X}_2 & \bar{X}_3 \\ d_1\bar{Y}_1 & d_1\bar{Y}_2 & d_1\bar{Y}_3 \\ d_2\bar{Z}_1 & d_2\bar{Z}_2 & d_2\bar{Z}_3 \end{pmatrix} A$$

that is

$$
\begin{pmatrix}
\bar{X}_1 & \bar{X}_2 & \bar{X}_3 \\
\bar{Y}_1 & \bar{Y}_2 & \bar{Y}_3 \\
\bar{Z}_1 & \bar{Z}_2 & \bar{Z}_3
\end{pmatrix}
=
\begin{pmatrix}
X_1 & X_2 & X_3 \\
d_1^{-1}Y_1 & d_1^{-1}Y_2 & d_1^{-1}Y_3 \\
d_2^{-1}Z_1 & d_2^{-1}Z_2 & d_2^{-1}Z_3
\end{pmatrix}
A^{-1}.
$$

He is able to state that the equivalence of Massau forms is a true equivalence relation; i.e., if each of two Massau forms is equivalent to a third then all three forms are equivalent to each other. It may be that a function has two Massau forms which are not equivalent, or indeed more than two such forms, and Warmus gives the following definitions.

A *uniquely nomographic function* is a function which has all its Massau forms equivalent in pairs.

A *doubly nomographic function* is one which has exactly two non-equivalent Massau forms.

A *K-nomographic function* for $K > 2$ is one which has exactly K non-equivalent Massau forms. From the definitions for the equivalence of Massau forms of $F(x, y, z)$, it is clear that the corresponding nomograms for $F(x, y, z) = 0$ of two such forms can be obtained from each other by homographic transformations. It is of interest to note that equivalent forms of

$$
\begin{vmatrix}
X_1 & X_2 & X_3 \\
Y_1 & Y_2 & Y_3 \\
Z_1 & Z_2 & Z_3
\end{vmatrix}
$$

can be obtained by,
   i. interchanging two columns and replacing signs of one row by their opposites
   ii. adding to one column a linear combination of the other columns
   iii. multiplying one row by non-zero a and one column by a^{-1}
   iv. multiplying one row (or column) by non-zero a and another row (or column) by a^{-1}.

Warmus is now able to develop theorems which give for functions of rank 2 or 3 the form of the Massau equivalent forms. These present us with no surprises but they do lead to a classification of the cases that can arise. By making preliminary assumptions, which cause little inconvenience and do not restrict the generality of the problem, Warmus is able to classify nomographic functions into seven different categories which he calls the *principal cases*.

One of the preliminary assumptions made is that for a nomographic function F there are only the following possible cases.

	F	G_1	G_2	G_3
1	$X_1G_1 + X_2G_2$	Y_1Z_1	Y_3Z_3	
2		$Y_1Z_1 + Y_2Z_2$		
3			$Y_3Z_3 + Y_4Z_4$	
4	$X_1G_1 + X_2G_2 + X_3G_3$	Y_1Z_1	Y_3Z_3	Y_5Z_5
5		$Y_1Z_1 + Y_2Z_2$		
6			$Y_3Z_3 + Y_4Z_4$	
7				$Y_5Z_5 + Y_6Z_6$

TABLE 4.1.

where the following are linearly independent in the pairs shown:

- (Y_1Y_2), (Y_3Y_4), (Y_5Y_6), (Z_1Z_2), (Z_3Z_4), and (Z_5Z_6)
- (X_1X_2) and (G_1G_2) in the first three cases
- $(X_1X_2X_3)$ and $(G_1G_2G_3)$ in the last four cases

The classification is done on the basis of the rank of $F(x, y, z)$ with respect to the variables x, y, and z and the ranks of the constituent functions $G_i(y, z)$. As an illustration, consider the first principal case $F = X_1G_1 + X_2G_2$ in which the function F is of rank 2 with respect to each of the variables x, y, and z. The functions G_1 and G_2 are both of rank 1.

For each principal case Warmus gives the associated principal Massau forms. For the first principal case $F = X_1 G_1 + X_2 G_2$, the principal Massau forms are

$$
\begin{vmatrix}
X_1 & X_2 & 0 \\
0 & Y_1 & Y_3 \\
Z_3 & 0 & Z_1
\end{vmatrix}
\quad \text{and} \quad
\begin{vmatrix}
X_1 & X_2 & 0 \\
Y_3 & 0 & Y_1 \\
0 & -Z_1 & -Z_3
\end{vmatrix}.
$$

Based on the proceeding, Warmus proves the following two theorems which he calls *fundamental theorems*. The m's, n's and r's in the second theorem are numbers which occur in the list of principal cases.

THEOREM. The function F is nomographic if, and only if, under the three preliminary assumptions one of the principal cases occurs.

THEOREM. If the function F is nomographic under the three preliminary assumptions then,

(a) it is doubly nomographic whenever the first principal case occurs, or the second principal case with $m_{31}n_{31} + m_{32}n_{32} \neq 0$, or the third principal case with $(r_{31} - r_{42})^2 + 4r_{32}r_{41} \neq 0$. In these cases every Massau form of the function F is equivalent to one of the two corresponding principal Massau forms, the two forms being non-equivalent.

(b) it is uniquely nomographic in the remaining cases and in the second principal case with $m_{31}n_{31} + m_{32}n_{32} = 0$ and the third principal case with $(r_{31} - r_{42})^2 + 4r_{32}r_{41} = 0$. In these cases every Massau form of the function F is equivalent to the corresponding principal Massau form.

As has already been noted, Warmus's efforts are directed to a classification scheme of nomographic functions based upon the concepts of rank and linear dependence; this classification incorporates the principal Massau forms associated with a nomographic function. Such a classification can give rise to an effective procedure which will determine a Massau form for a function when this is possible or indicate that the function is not nomographic. Warmus obtains such a procedure which he calls his *scheme of computations*.

There are nine computation schemes in Warmus's paper but they are not independent. They are used in the following manner.

Start with Scheme I. This has several exit points which either end the problem, showing that F is not nomographic, or pass to Scheme II. Scheme II, similarly, either terminates the problem in the case of a non-nomographic function or passes the computation to one of Schemes III, IV, V, VI, VII, VIII, or IX. Each of these latter schemes either will give Massau forms or indicate that the function is non-nomographic. The whole computation scheme is lengthy, occupying some twenty pages.

It cannot be claimed that Warmus has provided a simple solution to the problem, for the problem is not simple, yet he has done what earlier investigators did only incidentally; he has provided a method which leads the practical nomographer to a determinant form, when this is possible, and hence to a nomogram.

The illustrations given by Warmus show at the same time both that his scheme works and that it is extremely tedious. For example, the case,

$$F(x, y, z) = -2 - x^2 - y + z + x^2 y^2$$
$$+ e^x z - e^x y^2 z - x^2 \sqrt{(z)} + e^x \sqrt{z}(y + 2)$$

takes nine pages to arrive at the determinant form,

$$F(x, y, z) = \begin{vmatrix} -1 & x^2 & e^x \\ 1 - y^2 & 2 + y & 1 \\ -\sqrt{z} & z & 1 \end{vmatrix}$$

during the course of which four sets of values satisfying certain criteria have been found and fourteen 2×2 determinants evaluated. These determinants contain variables and are fairly complex with plenty of scope for error.

4.5. The Practical Approach of Džems-Levi

The final attempt we examine also dates from 1959 [**40**]. It appeared in an issue of the Russian journal *Computational Mathematics* which was entirely devoted to nomography. The author, G.E. Džems-Levi, approached the problem of anamorphosis in a manner reflecting the times, times in which the development of computational aids was making approximation techniques more readily acceptable. In fact, Džems-Levi's approach is based on a method of constructing nomograms which is an approximation method. It will be necessary to describe this method, but first some preliminary remarks.

In his paper Džems-Levi assumes that the given equation is nomographible and that only the scales have to be found; i.e., the f_i and g_i of the determinant

$$\begin{vmatrix} f_1(x) & g_1(x) & 1 \\ f_2(y) & g_2(y) & 1 \\ f_3(z) & g_3(z) & 1 \end{vmatrix}.$$

However, the arguments apply equally to non-nomographible equations, the difference being that nomographible equations are accurately portrayed while non-nomographible equations lead to approximate nomograms. An interesting aside is his claim that the method can be applied to problems on the feasibility of representing a given equation in n variables in the form of a superposition of functions of a smaller number of variables.

The approximation method referred to is due to Gorodskii and appears to date from 1939 [**48**]. Applied to a relation of the form $z = F(x, y)$, the steps are as follows:

   i. An arbitrary but convenient scale for z is chosen. It must be monotonic and continuous.

  ii. Two arbitrary points of the y scale, y_1 and y_2, are selected.

 iii. A series of values of z are computed from $z = F(x, y)$ using the points y_1 and y_2 and a series of values for x.

 iv. Next, the x's are plotted; x_i is the intersection of the line joining y_1 to $F(x_i, y_1)$ with the line joining y_2 to $F(x_i, y_2)$, (Figure 4.1).

  v. When the scale of x has been constructed the scale of y may be constructed using the points x_1 and x_2 as pivots.

 vi. The nomogram can now be improved. Through pairs of points of two scales a series of lines is drawn to give one and the same value of a third variable. If the nomogram is accurate, then all the lines will pass through one point; more usually they will describe a region in which case the scale is reconstructed with the value assigned to the centroid of the region. This latter process may be accomplished numerically. The whole process is now repeated for the other two scales.

The first five steps of the process are illustrated in Figure 4.2.

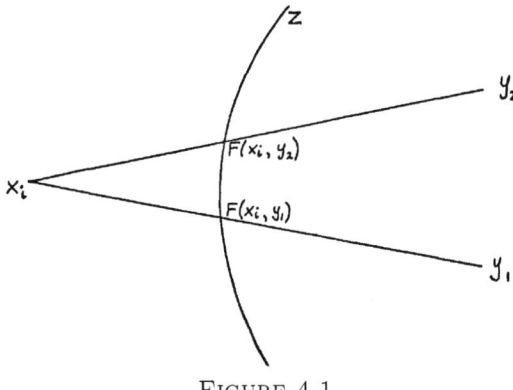

FIGURE 4.1

Figure 4.2 is a partially constructed nomogram for $z = xy$ which illustrates Gorodskii's method. The z scale was chosen to be rectilinear and logarithmic. The two y pivots, y_1 and y_2, were chosen to be on a straight line parallel to the z scale; y_1 was assigned the value $y = 1$ and y_2 the value $y = 4$. The points 1, 2, 3, 4, 5, and 6 were located for the x scale by the rays emanating from y_1 and y_2.

To construct the y scale the pivots x_1 and x_2 were selected at the points $x = 1$ and $x = 3$. The points 2, 3, 5, and 6 of the y scale were located by the broken rays emanating from x_1 and x_2.

It will be noted that both the x and y scales are rectilinear; they are also logarithmic as one would expect them to be.

Returning to the main theme, we note that if the resulting nomogram corresponds exactly to $z = F(x, y)$ and y_1 and y_2 are where they should be then the construction method is exact. The reasoning may then be expressed in an analytical form. Džems-Levi's approach stems from the proposition that if one scale of a nomogram is known, then the determination of the other scales can be carried out in an analytical manner.

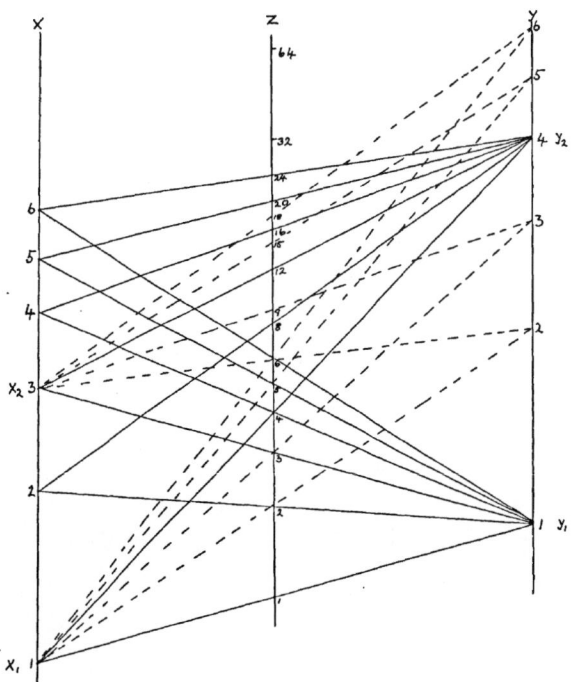

FIGURE 4.2. Gorodskii's method applied to $z = xy$

Consider $z = F(x, y)$ where

$$\frac{\partial F}{\partial x} \cdot \frac{\partial F}{\partial y} \neq 0$$

in the region G of the xy plane under consideration. A nomogram is supposed constructed, if that is possible, using fixed points as described above. The y scale is to be curved in the sense that there is a part of it on which no three points are in a straight line. It is this part which is considered. A projective transformation changes four points on the y scale into the following points,

$$y_1(0, \infty), \quad y_2(-\infty, 0), \quad y_3(0, 0), \quad \text{and} \quad y_4(1, 1).$$

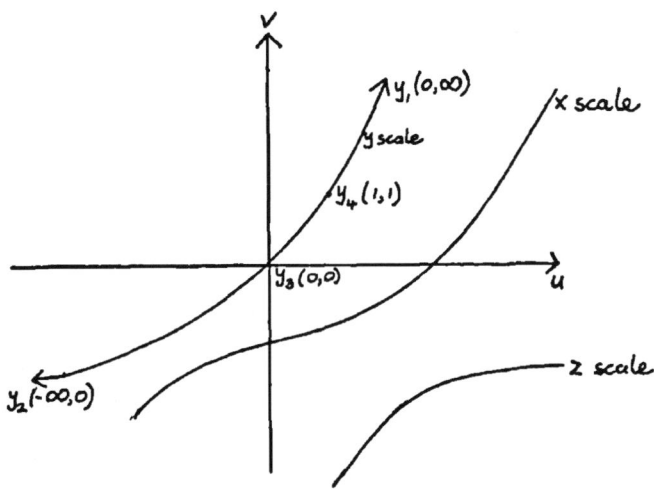

FIGURE 4.3

Suppose that after this projective transformation the equation of the nomogram can be expressed in the form,

$$\begin{vmatrix} f_1(x) & g_1(x) & 1 \\ f_2(y) & g_2(y) & 1 \\ f_3(z) & g_3(z) & 1 \end{vmatrix} = 0$$

in which the coordinates of the transformed system, u and v, are given by $u = g_i$ and $v = f_i$. Figure 4.3 portrays the new situation.

In Figure 4.4, x is an arbitrary point on the x scale. The lines drawn through x in the following four directions:

    i. parallel to the u axis
    ii. parallel to the v axis
    iii. through the origin
    iv. through the point $(1, 1)$

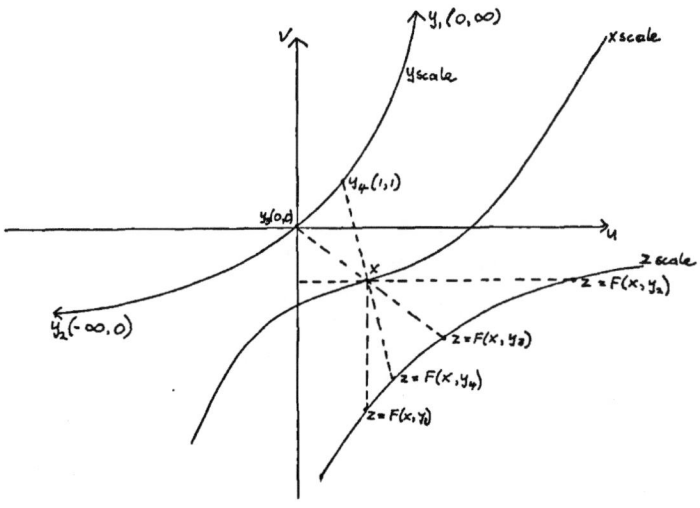

FIGURE 4.4

must intersect the z scale at the following points,

$$z = F(x, y_2), \quad z = F(x, y_1), \quad z = F(x, y_3), \quad \text{and} \quad z = F(x, y_4)$$

respectively.

It follows that, from the u and v coordinates of x we must have

$$g_1(x) = g_3(F(x, y_1)) \tag{4.23}$$

and

$$f_1(x) = f_3(F(x, y_2)). \tag{4.24}$$

By considering the slopes of the lines we have,

$$\frac{f_1(x)}{g_1(x)} = \frac{f_3(F(x, y_3))}{g_3(F(x, y_3))} \tag{4.25}$$

for the line joining x to the origin, and

$$\frac{f_1(x) - 1}{g_1(x) - 1} = \frac{f_3(F(x, y_4)) - 1}{g_3(F(x, y_4)) - 1} \tag{4.26}$$

for the line joining x to $(1, 1)$.

Džems-Levi's proposition states that one scale is to be known and we assume that this is the z scale. Since y_1 and y_2 are known and x is arbitrary, (4.23) and (4.24) will give the x scale; i.e., $g_1(x)$ and $f_1(x)$.

To obtain the y scale fix two points on the x scale, $x = x_1$ and $x = x_2$. Using the determinant form we can write,

$$\begin{vmatrix} f_1(x_1) & g_1(x_1) & 1 \\ f_2(y) & g_2(y) & 1 \\ f_3(F(x_1,y)) & g_3(F(x_1,y)) & 1 \end{vmatrix} = 0$$

and

$$\begin{vmatrix} f_1(x_2) & g_1(x_2) & 1 \\ f_2(y) & g_2(y) & 1 \\ f_3(F(x_2,y)) & g_3(F(x_2,y)) & 1 \end{vmatrix} = 0$$

from which, by suitable algebraic manipulation, expressions for $f_2(y)$ and $g_2(y)$ can be obtained. Thus, the knowledge of one scale, in this case that of z, and two points, namely y_1 and y_2, are sufficient to determine the other two scales.

It will be noted that (4.25) and (4.26) were not used in the proceeding argument. If they are used in conjunction with (4.23) and (4.24) one can obtain,

$$\frac{f_3(F(x,y_2))}{g_3(F(x,y_1))} = \frac{f_3(F(x,y_3))}{g_3(F(x,y_3))}$$

and

$$\frac{f_3(F(x,y_2)) - 1}{g_3(F(x,y_1)) - 1} = \frac{f_3(F(x,y_4)) - 1}{g_3(F(x,y_4)) - 1}$$

from which the functions $g_3(z)$ and $f_3(z)$ can be determined.

Džems-Levi remarks that the solution of functional equations, such as those which arise above, has not been investigated to any marked degree. He therefore proposes the substitution of differential equations for functional equations.

To obtain the differential equations, his approach is as follows. Given

$$z = F(x,y) \tag{4.27}$$

he postulates the existence of a normal nomogram constructed from fixed points, as already described, and having a determinant representation,

$$\begin{vmatrix} g_1(x) & f_1(x) & 1 \\ g_2(y) & f_2(y) & 1 \\ g_3(z) & f_3(z) & 1 \end{vmatrix} = 0. \tag{4.28}$$

A nomogram is said to be *normal* if it can be constructed and used as a computational instrument. y is considered as a function of x and z and it is required that $\frac{\partial y}{\partial z}$ obtained from (4.27) and from (4.28) should be equal. A corresponding result must also apply to $\frac{\partial^2 y}{\partial z^2}$.

The resulting expressions contain $f_3(z)$ which, when eliminated, produces the identity,

$$\frac{N}{p} = \frac{\Delta_1 x}{\Delta_1} - \frac{2\Delta_2 x}{\Delta_2} + p\left(-\frac{\Delta_2 y}{\Delta_2} - \frac{2\Delta_1 y}{\Delta_1}\right) \tag{4.29}$$

in which

$$N(x,y) = -\frac{\partial^2 y}{\partial x^2} \quad \text{and} \quad p(x,y) = -\frac{\partial y}{\partial x}$$

and

$$\Delta_1 = f_1'(g_2 - g_1) - g_1'(f_2 - f_1) \quad \text{and} \quad \Delta_2 = f_2'(g_2 - g_1) - g_2'(f_2 - f_1).$$

Džems-Levi states that the expression,

$$\frac{\Delta_1 x}{\Delta_1} - \frac{2\Delta_2 x}{\Delta_2}$$

contained in (4.29) is none other than Grönwall's C while

$$-\frac{\Delta_2 y}{\Delta_2} - \frac{2\Delta_1 y}{\Delta_1}$$

is Grönwall's D. It is then a simple matter to show that equation (4.29) is Džems-Levi's equivalent of Grönwall's $D = MC + N$ (4.7).

Grönwall and Džems-Levi are not the only two investigators to have arrived at this point, for Džems-Levi notes that I.A. Vilner [133], and S.V. Smirnov [125], had also arrived there. However, their subsequent treatments diverge.

The Džems-Levi approach is to convert (4.29) into a differential equation in x by substituting for y some value y_0 within the permitted range. A second differential equation is obtained by differentiating (4.29) with respect to y and making again the substitution $y = y_0$.

This system of two differential equations contains eight unknown constants, namely

$$g_2(y_0), \quad g_2'(y_0), \quad g_2''(y_0), \quad g_2'''(y_0),$$

and

$$f_2(y_0), \quad f_2'(y_0), \quad f_2''(y_0), \quad f_2'''(y_0).$$

However, if the y scale; i.e., $u = g_2(y)$ and $v = f_2(y)$, has non-zero curvature at $y = y_0$, then a projective transformation can be found such that, at the corresponding point on the transformed scaled denoted by $\bar{g}_2(y_0)$ and $\bar{f}_2(y_0)$, the constants are given by,

$$\bar{g}_2(y_0), \quad \bar{g}_2'(y_0), \quad \bar{g}_2''(y_0), \quad \bar{g}_2'''(y_0),$$

and

$$\bar{f}_2(y_0), \quad \bar{f}_2'(y_0), \quad \bar{f}_2''(y_0), \quad \bar{f}_2'''(y_0).$$

A further simplification of the two differential equations is possible if the x scale is assumed to be curvilinear such that u and v can be expressed as

$$u = \frac{f_1}{g_1} \quad \text{and} \quad v = \frac{2}{g_1}.$$

The resulting differential equations are then,

$$u'' = u' \left(\frac{N}{p} + p\,u \right) - v'p \tag{4.30}$$

and

$$v'' = 4u'^2 + u' \left(3vp - 2u^2p - 2p_y u - 2 \left(\frac{N}{p} \right)_y \right) + v' \left(\frac{N}{p} - p\,u + 2p_y \right) \tag{4.31}$$

Provided that $p(x, y_0) \neq 0$, equations (4.30) and (4.31) will eventually give the x scale, since,

$$g_1(x) = \frac{2}{v} \quad \text{and} \quad f_1(x) = \frac{2u}{v}.$$

Of course, the solutions will contain four constants of integration.

Having found the x scale, the remaining scales present no difficulty. Taking the equation (4.27); i.e., $z = F(x, y)$, it is rearranged to give

$$x = \psi(y, z). \tag{4.32}$$

The relationship

$$p(x, y) = \frac{f_3(z) - f_2(y)}{f_1(x) - f_3(z)} \cdot \frac{\Delta_1}{\Delta_2}, \tag{4.33}$$

where Δ_1 and Δ_2 are as defined in (4.29), is obtained by differentiating (4.28) partially with respect to x.

By eliminating x between (4.32) and (4.33), an identity in y and z is obtained. Putting $y = y_0$ into this identity and rearranging it we get,

$$f_3(z) = \frac{p\,g_1 f_1}{p\,g_1 + g_1 f_1' - g_1' f_1}$$

in which

$$x = \psi(y_0, z).$$

Similarly, by putting $x = \psi(y, z)$ in (4.28) an identity is obtained from which $g_3(z)$ may be determined. We get

$$g_3(z) = f_3(z) \frac{g_1(\psi(y_0, z))}{f_1(\psi(y_0, z))}.$$

To obtain $g_2(y)$ and $f_2(y)$ we take two arbitrary values of z, z_1 and z_2, and find the corresponding values of $f_3(z)$ and $g_3(z)$. From the determinant form (4.28), after replacing x by $\psi(y, z)$ and in turn z by z_1 and z by z_2, two linear equations are obtained from which $g_2(y)$ and $f_2(y)$ may be found. It will be seen from Figure 4.5 that this is an application of Gorodskii's method.

It still remains to determine the constants of integration. This can be done by choosing four sets of three values which satisfy $z = F(x, y)$ and substituting them into the determinant form obtained by the method described above. This will provide four equations in the four unknown constants.

Expanding the determinant (4.28) along the second row and replacing x by $\psi(y, z)$ gives,

$$g_2(y)\,(f_1(\psi(y, z)) - f_3(z)) - f_2(y)\,(g_1(\psi(y, z)) - g_3(z)) + \tag{4.34}$$
$$g_1(\psi(y, z)) f_3(z) - g_3(z) f_1(\psi(y, z)) = 0.$$

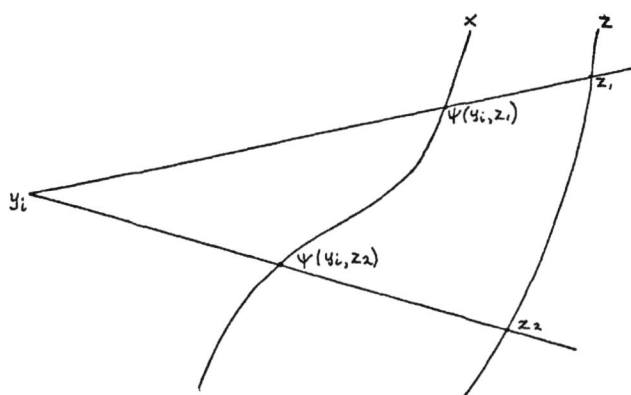

FIGURE 4.5

z_1 and z_2 satisfy equation (4.34) and hence, regarding it as an equation in $g_2(y)$ and $f_2(y)$, we can obtain two more such equations by the successive substitutions $z = z_1$ and $z = z_2$. For consistency the following identity must hold,

$$\begin{vmatrix} f_1(\psi(y,z_1)) & g_1(\psi(y,z_1)) & g_1(\psi(y,z_1))f_3(z_1) \\ -f_3(z_1) & -g_3(z_1) & -f_1(\psi(y,z_1))g_3(z_1) \\ f_1(\psi(y,z_2)) & g_1(\psi(y,z_2)) & g_1(\psi(y,z_2))f_3(z_2) \\ -f_3(z_2) & -g_3(z_2) & -f_1(\psi(y,z_2))g_3(z_2) \\ f_1(\psi(y,z)) & g_1(\psi(y,z)) & g_1(\psi(y,z))f_3(z) \\ -f_3(z) & -g_3(z) & -f_1(\psi(y,z))g_3(z) \end{vmatrix} = 0 \qquad (4.35)$$

Nothing in the process described above is restricted by the nomogrammibility of the equation $z = F(x, y)$. The process may be carried out whether the result will, turn out to be an exact nomogram or an approximate one. In the terms of Džems-Levi's investigation the problem with which this chapter is concerned is that of the existence of an exact alignment nomogram for $z = F(x, y)$.

The existence of such a nomogram is determined in this case by the reduction of (4.35) to an identity when the values found for the constants of integration have been inserted. Enough of Džems-Levi's paper has been considered for the purpose of this investigation into the theoretical problem. However, it seems appropriate to note briefly the substance of the rest of the paper.

The application of the method to the joint nomogram of two equations,

$$z = F(x, y) \quad \text{and} \quad \bar{z} = G(x, y)$$

in which the x and y scales coincide, yields particularly simple results. The method is also applied in some detail to equations of the third nomographic order; i.e., $Z(z) = X(x) + Y(y)$, and to the fifth order in the form,

$$f_1(x) = \frac{f_2(y) + f_3(z)}{g_2(y) + g_3(z)}$$

in which the x scale is linear, and also to a system of equations containing linear scales.

A result obtained for the third nomographic order in the form of the equation $z = x + y$ is worth repeating. From $z = x + y$ we have, regarding y as a function of x and z,

$$p = -\frac{\partial y}{\partial x} = 1 \quad \text{and} \quad N = \frac{\partial^2 y}{\partial x^2} = 0.$$

For a linear scale of x we can take $u = f_1$ and $v = 2$. (In (4.30) and (4.31) where $u = f_1 g_1^{-1}$ and $v = 2g_1^{-1}$ we set $g_1 = 1$.) Equation (4.30),

$$u'' = u' \left(\frac{N}{p} + pu \right) - v'p,$$

now becomes

$$f_1'' = f_1 f_1'$$

and therefore

$$x = 2 \int \frac{df_1}{f_1^2 + c_1} + c_2.$$

The constant c_1 may be smaller than, equal to, or greater than zero and, according to which, and with an appropriate transform of $z = x + y$ into

$az + b + c = (ax + b) + (ay + c)$, Džems-Levi shows that there are three, and only three, projectively distinct scales for x. They are,

$$f_1(x) = e^{ax},$$
$$f_1(x) = ax,$$
$$\text{and} \quad f_1(x) = \tan(ax)$$

and all other scales may be obtained from them by projective transformations. Both Grönwall and Warmus obtained results in this vein.

As a result of obtaining a canonical form for equations of the third nomographic order, Džems-Levi is able to produce a result which is a particular form of Saint-Robert's criterion. Equations of the fourth order are not dealt with since these are the subject of a separate paper [**37**].

His final contribution to the problem of anamorphosis is the replacement of the necessity to solve differential equations by the requirement to solve a system of algebraic equations. Returning to the identity (4.29), we note that instead of regarding it as a differential equation in $f_1(x)$ and $g_1(x)$ it may be regarded as an ordinary equation in f_1, f_1', f_1'', g_1, g_1', and g_1''. By differentiating both sides with respect to y we noted that a second differential equation could be obtained which, with $y = y_0$, gave a second equation. This may now be repeated until we have six equations; i.e., by taking up to the fifth derivative of both sides of (4.29). This means that there will be a seventh derivative from $F(x, y)$. The system of equations, now considered an algebraic system, will contain 16 constants which are,

$$g_2(y_0), \quad g_2'(y_0), \quad \ldots, \quad g_2^v(y_0),$$

and

$$f_2(y_0), \quad f_2'(y_0), \quad \ldots, \quad f_2^v(y_0).$$

These may be reduced to eight by the use of the projective transformation hinted at earlier. The remaining eight constants are found by ensuring conformity between the system of equations and the solutions. This is done by equating expressions for, in one case, $g_1(x)$ and in another $f_1(x)$. The two expressions for $f_1(x)$, for example, are obtained in the following ways; as a result of the solution of the algebraic system and, by differentiation with respect to x of equations in the system and the elimination of the derivative of $f_1(x)$.

An example for the case

$$f_1(x) = \frac{f_2(y) + f_3(z)}{g_2(y) + g_3(z)}$$

will illustrate. The algebraic system arrived at with $y = y_0$ is

$$\frac{N}{p} = \frac{f_1''}{f_1'} - pf_1 \tag{4.36}$$

$$\left(\frac{N}{p}\right)_y = 2f_1' - p_y f_1 - p(f_1^2 + af_1 - b) \tag{4.37}$$

$$\left(\frac{N}{p}\right)_{yy} = 2f_1'(2f_1 + a) - p_{yy} f_1 - 2p_y(f_1^2 - af_1 - b) - \tag{4.38}$$

$$p\left(2f_1^3 + 3af_1^2 - f_1(c - 3b) - d\right) \tag{4.39}$$

in which

$$a = g_2'''(y_0),$$
$$b = f_2'''(y_0),$$
$$c = g_2''''(y_0),$$
$$\text{and} \quad d = f_2''''(y_0).$$

The system gives,

$$f_1(x) = \frac{\left(\frac{N}{p}\right)_{yy} - \left(\frac{N}{p}\right)_y a - 2p_y\, b + p\,(ab - d)}{2\left(\frac{N}{p}\right)_y - p_{yy} - ap_y + p\,(b - c + a^2)}. \tag{4.40}$$

By differentiating (4.37) with respect to x and eliminating f_1'' between the result and (4.36), an equation in f_1' is obtained. Eliminating f_1' between this new equation and (4.39), an equation in f_1 is obtained. By eliminating f_1 between this new equation and (4.40) an identity results from which $a, b, c,$ and d may be found. However, if the equation is of the third order; i.e., of the form

$$f_3(z) = f_1(x) + f_2(y),$$

the method fails.

4.6. Conclusion

Any comparison made between the four papers described in this section would be rather artificial. One reason for this is that, in part, a person's position on any issue is determined by his philosophical views. The philosophical positions of the authors are by no means the same.

Grönwall has presented a paper which is in the classical mathematical tradition, his main conclusion is based upon the existence of the common solution C of two partial differential equations. But an object may exist without it being known to exist and there are those who would object to an existence proof.

Kellogg's paper seems to me to present a more readily acceptable solution in that his matrix N can be formed and its rank determined by well established methods but the actual execution may be tedious and complicated. However, Kellogg's paper served one very useful purpose; it appears to have given an idea to Warmus.

Warmus has produced a computation scheme based upon a classification system; it is thorough, contains interesting mathematical ideas and in any given case will either indicate that no nomogram is possible or will give the appropriate determinant forms. It is the paper that should be consulted by anyone wishing to construct a nomogram of some complexity.

Džems-Levi was clearly influenced by Grönwall's paper and indeed he is at pains to point out similarities between his results and Grönwall's when these occur. However, their papers reach a common point and then go in very different directions. The philosophy behind Džems-Levi's approach is one which would find favor with the Intuitionist School for it is a constructive proof; at a very basic level it seems to be saying that a nomogram exists if it can be constructed, though I accept that this is something of an oversimplification. I must declare my own leaning towards the Džems-Levi approach.

CHAPTER 5

Later Developments

5.1. Russian Advances

During the 1950's the Russian interest in nomography was strong. The work of G.E. Džems-Levi on anamorphosis dates from that period. Džems-Levi was just one of the contributors to an issue of *Vychislitelnaya Matematika*, (Computational Mathematics), which was devoted entirely to nomography. The other contributions are examined here. In general they represent a modern approach to the subject which reflects changing attitudes in mathematics. For example, consideration of approximate nomograms seems to be a reflection of the contemporaneous development in numerical analysis.

D.C. Lapteva considers a particular aspect of the projective transformation of alignment nomograms in which the resulting scale of the unknown variable is rectilinear [**73**]. The particular aspect which interests him concerns the errors present in results obtained from such nomograms. The manner in which the error in the solution is estimated will determine the form of the scale which carries the solution, in particular, if the error estimate is measured in terms of relative error then, for a given size of scale, the greatest accuracy is given by a logarithmic scale, as shown in Appendix C. Even for a nomogram having a uniform solution scale it may be that the problem will be better served if the scale is logarithmic.

The particular projective transformation considered for this purpose is a *homology*. A homology is a transformation which leaves invariant every point on a given line w and every line through a given point P, where P is not on the line w. Lapteva writes this transformation as,

$$x_1 = \frac{bx}{y+b} \tag{5.1}$$

$$\text{and} \quad y_1 = \frac{(L+b)y}{y+b}$$

where x and y are the old coordinates, x_1 and y_1 the new coordinates, L some fixed quantity and b the transformation parameter.

If $x_1 = x$, then $y = 0$, showing that the fixed line is the x axis. If $x = 0$ and $y = L$, then $x_1 = 0$ and $y_1 = L$ showing that the point $(0, L)$ is a fixed point.

If $y = \infty$ for some arbitrary x value a, then $x_1 = 0$ and $y_1 = L + b$, showing that straight lines parallel to the y axis are transformed into straight lines intersecting at the point $(0, L + b)$.

He selects a system of coordinates which makes the solution scale coincide with the y axis. L is taken to be the length of the solution scale; the transformation (5.1) will therefore leave it unchanged. The transformed nomogram can now be constructed without calculation. Consider the transformation of one of the scales, α.

In Figure 5.1, M is an arbitrary point on the scale α and M_x its projection on the x axis. The straight line of which MM_x forms part will be transformed by (5.1) into the straight line k passing through M_x, and B, where B is the point with coordinates $(0, L+b)$. The point M is then transformed into a point M' on the line k. M' will also lie on a straight line l passing through M and the point $A = (0, L)$, since lines through A are unchanged by the transformation. It is easy to construct the whole transformed scale of α in this way. If the original scale of α is rectilinear and parallel to the y axis, the construction is greatly simplified.

The greatest problem in this transformation is the choice of the parameter b. An analytical method by M.V. Pentkovsky exists but this leads to long and complicated calculations and furthermore examines only one scale and not the nomogram as a whole [110]. In practice it is less important to know the exact value of b than the range of values for which the scale more or less approximates to the given form. The interesting suggestion here is that another nomogram

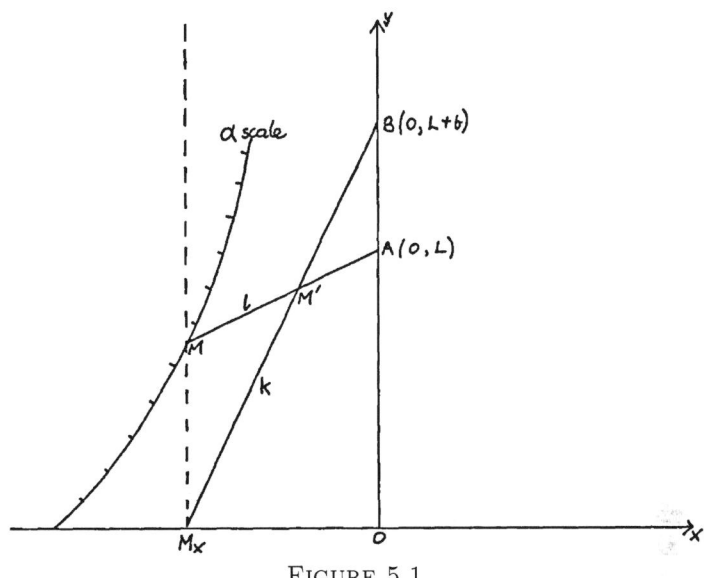

FIGURE 5.1

should be constructed for the second equation of (5.1) to show the influence of change in b on the transformed scale. Džems-Levi is credited with such a nomogram which is of the intersection form [33]. However, Lapteva prefers a different approach using an alignment nomogram.

The equation

$$y_1 = \frac{(L+b)y}{y+b}$$

is written as

$$\log \frac{L-y_1}{y_1} = \log \frac{L-y}{y} + \log \frac{b}{L+b}$$

an equation of the third nomographic order.

A nomogram is constructed having the scales y and b on an ellipse with a uniform scale for y_1 along the major axis. To construct this, use has been made of a skeleton nomogram classified as 321 by Pentkovsky [111]. This is reproduced as Figure 5.2.

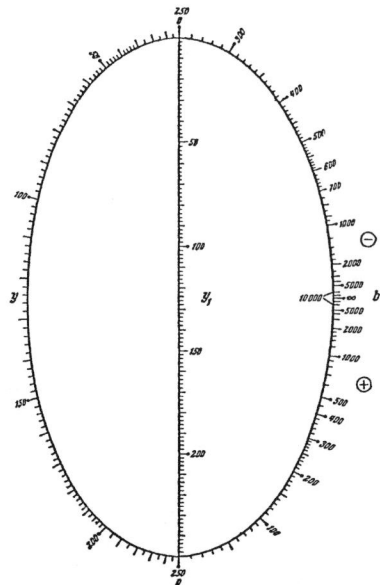

FIGURE 5.2. Pentkovsky's Skeleton Nomogram

To find a suitable range of values of b which will transform the y scale into an approximate logarithmic scale, the following procedure is used. The nomogram of Figure 5.2 is itself used as a skeleton. Selected rounded values of the variable y are marked on the y arc, due account having been taken of the range of values appropriate to the particular problem. On y_1 the logarithmic values of the same range of values are plotted. Alignment is made between the rounded values on y with the corresponding point on the logarithmic scale on y_1. The result is that a range of values of b is determined on the third scale. If the conditions of the problem require, for example, an elongation of a certain point of the scale, then such consideration can lead to a best value from within the range.

In an interesting but rather short paper I.N. Denisyuk considers the construction of empirical formulae for data which are believed to approximate to a straight line on a logarithmic or half-logarithmic base [23]. He is able to give formulæ which enable a relationship of the form

$$y = a + \frac{b}{x + c}$$

to be obtained without any great difficulty. His method is based upon the fact that within certain limits, and with a certain choice of three points on the x scale, the transformation,

$$X = \frac{L}{1 + x(x_1 x_2)^{-\frac{1}{2}}}$$

will lead to a scale little different from the logarithmic one. In fact it is the scale deviating least in Chebyshev's sense. In place of a logarithmic base he considers a base constructed from two projective transformations

$$\frac{1}{1 + x(x_1 x_2)^{-\frac{1}{2}}} \quad \text{and} \quad \frac{1}{1 + y(y_1 y_2)^{-\frac{1}{2}}}$$

along the axes. Here (x_1, y_1) and (x_2, y_2) are the coordinates of two points on the supposed line, chosen so that they are at distances from the ends of the section under consideration of approximately one fifth the width of the transformed section. Throughout it is assumed that $x_1 < x_2$.

Denisyuk's results may be summarized as follows. If

$$x_1 = \alpha x_2 \quad \text{and} \quad y_1 = \beta y_2,$$

then two constants a and b are found as follows,

$$a = \frac{\sqrt{\alpha\beta} - 1}{\sqrt{\alpha} - \sqrt{\beta}} \quad \text{and} \quad b = \frac{(1 - \alpha)(1 - \beta)}{(\sqrt{\alpha} - \sqrt{\beta})^2}.$$

Using these, the empirical formula is obtained from

$$y = y_2 \sqrt{\beta} \left(a - \frac{b}{x(x_2\sqrt{\alpha})^{-1} + a} \right).$$

He gives an example to illustrate. If $x_1 = 0.5$, $x_2 = 2$, $y_1 = 0.8$, and $y_2 = 5$, then $\alpha = 0.25$, $\beta = 0.16$, $a = -8$, and $b = 63$ leading to

$$y = \left(\frac{126}{8 - x} - 16 \right).$$

In the case in which the y axis is uniform and only the x axis is to approximate to a logarithmic scale, the empirical formula is given by,

$$y = \frac{y_2}{1 - \sqrt{\alpha}} \left(1 - \beta\sqrt{\alpha} + \frac{(1 + \sqrt{\alpha})(\beta - 1)}{1 + x(x_2\sqrt{\alpha})^{-1}} \right).$$

The material in this paper is clearly based upon profound work only hinted at in the paper. The author adds the comment that some of the material had been presented at the seminar on nomography at Moscow State University in April 1955 and that further material had been presented at a seminar at the Computer Centre of A.N., USSR on May 23, 1957.

V.A. Cherpasov and G. E. Džems-Levi present a method for the calculation of approximate alignment nomograms using a computer [16]. The method is essentially one of finding an optimal group of parameters by successive approximations. A logical diagram for the process is given and although the form is unfamiliar it can be rewritten as a Western-type flow chart with little difficulty.

I.N. Denisyuk, in a second paper, describes a nomogram for the construction of some polynomials [24]. In particular he is concerned with calculating generalized Laguerre polynomials $L_n(t, \lambda)$ which he approaches through the solution of a boundary value difference equation. From his references we can infer that Denisyuk's work on Laguerre polynomials dates from the early 1950's. It is of some interest to note that during 1950 one of the last users of the Manchester University Mark I computer before it was dismantled was Dr. D.G. Prinz of Ferrenti Ltd. who computed Laguerre functions in connection with the control of guided weapons [75]. Whether or not the Russian calculations arose from guided weapon technology the different approaches to calculation give rise to speculation about Russian computer development at this time and perhaps indicate why Russian interest in nomography was so great.

The use of a transparency as part of a nomogram has a long history. Lallemand's hexagonal nomogram incorporated one (see Chapter 2, Section 5), and the concept was further developed by Margoulis in the early 1920's ([80] and [81]). The method receives a full treatment in a long and detailed paper by G.S. Khovanskii [60]. A brief outline of his ideas follows.

The relationship
$$f_3 = F(f_{12} + f_4, g_{12} + g_4)$$

where $f_{ij} = f(\alpha_i, \alpha_j)$ in the four variables α_1, α_2, α_3, and α_4 can be represented by a nomogram with an orientated transparency. The result has the advantage that the families of curves in α_1, and α_2, which would normally form a binomial field, can be separated.

Two auxiliary variables are introduced,

$$M = f_{12} + f_4 \quad \text{and} \quad N = g_{12} + g_4$$

giving the relationship $f_3 = F(M, N)$. These are written,

$$f_{12} - 0 = f_{12} - D = M - f_4$$
$$\text{and} \quad g_{12} - 0 = g_{12} - D = N - g_4.$$

This strange form is used because it is intended that in the left hand parts α_2 is an auxiliary variable while in the center parts it is α_1 which is the auxiliary variable. The zeros indicate positions which in this example are not filled but which may be filled in other cases.

The following tables show the elements of the nomogram when scale factors μx, μy, δx, and δy have been introduced, $(a_0, a_0', $ etc. are constants suitably chosen).

	α_1 Lines (α_2 aux.)	α_2 Lines (α_1 aux.)	MN Field
x	$a_0 +$ $\mu x(f_{12} + \delta x g_{12})$	$a_0 + a +$ $\mu x(f_{12} + \delta x g_{12})$	$a_0 + c +$ $\mu x(M + \delta x N)$
y	$b_0 +$ $\mu y(\delta y f_{12} + g_{12})$	$b_0 + b +$ $\mu y(\delta y f_{12} + g_{12})$	$b_0 + d +$ $\mu y(\delta y M + N)$

TABLE 5.1. Base

The base will contain three sets of curves. The construction of the sets for a α_1 and a α_2 is obvious. The α_3 set is obtained by plotting on the MN field according to the relationship $f_3 = F(M, N)$. Also on the base it is necessary to have a set of parallel straight lines to facilitate the orientation of the transparency. The transparency will have two fixed points, A_1 and A_2, a scale for α_4 and an orientating straight line. The form of the base and transparency are illustrated in Figure 5.3.

	Fixed Point A_1	Fixed Point A_2	f_4 and g_4 Scales
x	a_0'	$a_0' + a$	$a_0' + c + \mu x(f_4 + \delta x g_4)$
y	b_0'	$b_0' + b$	$b_0' + d + \mu y(\delta y f_4 + g_4)$

TABLE 5.2. Transparency

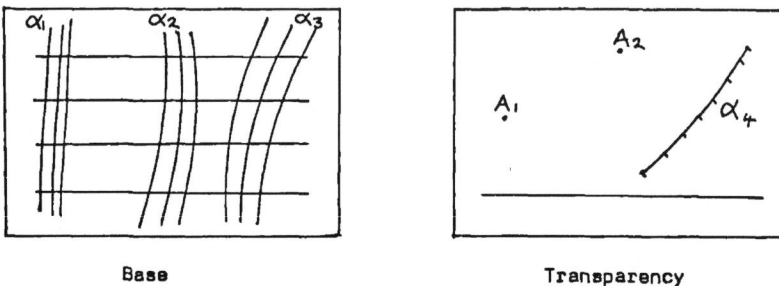

Base   Transparency

FIGURE 5.3. Base and Transparency

The use of the nomogram is as follows. With the correct orientation, which is crucial, A_1 is made to coincide with the value of α_1 and A_2 with that of α_2. Then the unknown value, say α_3, is that value which corresponds to the given value of α_4. Khovanskii writes this as,

$$A_1 \vdash\dashv \alpha_1, \quad A_2 \vdash\dashv \alpha_2, \quad \text{and} \quad \alpha_4 \vdash\dashv \alpha_3$$

a notation very similar to that used by d'Ocagne in his *Traité*. The construction of such a nomogram is more difficult than it may seem from this brief description for considerable care is required over the choice of the scale factors if reasonable accuracy is to be obtained.

Khovanskii treats in a similar manner the forms,

$$f_4 + F(\alpha_3, g_{12}) + f_{12} = 0,$$

$$f_{12} + g_{12}g_{34} + f_{34} = 0,$$

$$\begin{vmatrix} f_1 & g_1 & 1 \\ f_2 & g_2 & 1 \\ f_{34} & g_{34} & 1 \end{vmatrix} = 0,$$

$$f_1 = \frac{f_2 + f_{34}}{g_2 + g_{34}},$$

$$f_1 f_2 f_{34} + (f_1 + f_2)g_{34} + h_{34} = 0,$$

and $\quad f_1 f_{34} + f_2 g_{34} + h_{34} = 0.$

A similar treatment is given for the general case

$$\begin{vmatrix} f_{12} & g_{12} & 1 \\ f_{34} & g_{34} & 1 \\ f_{56} & g_{56} & 1 \end{vmatrix} = 0 \tag{5.2}$$

which is obtained by eliminating γ, β, and δ from

$$f_{12} - \beta = f_{34} - \gamma = f_{56} - \delta$$

and

$$\log g_{12} - \log \beta = \log g_{34} - \log \gamma = \log g_{56} - \log \delta.$$

The transparency carries the curve $y = \log x$ and the base has three binomial fields each of which is obtained by taking for the x coordinate the appropriate element of the first column and for the y coordinate the logarithm of the corresponding element of the second column.

$$\begin{vmatrix} f_1 & f_4 & 1 \\ f_2 & f_5 & 1 \\ f_3 & f_6 & 1 \end{vmatrix} = 0.$$

The nomogram is particularly simple as the binomial fields consist of orthogonal straight lines. The families of parallel straight lines α_4, α_5, and α_6 are also used as orientating guide lines (Figure 5.4). A wide range of other variants of (5.2) is also examined.

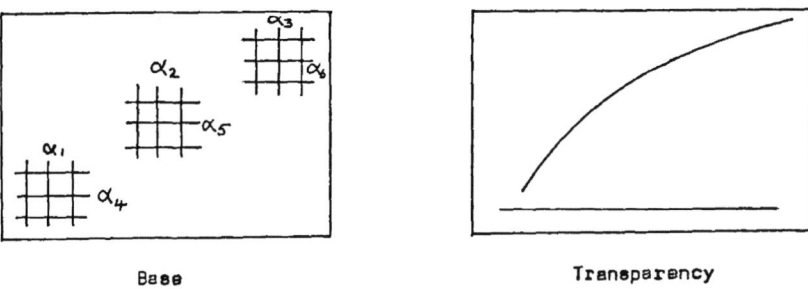

Base   Transparency

FIGURE 5.4. Base and Transparency

A particularly interesting aspect of Khovanskii's work concerns nomographic methods for the approximate representation of a function of one variable. In this work the transparency is allowed three degrees of freedom. These methods follow from a consideration of the relationships,

$$\phi(v, f(u)\cos\alpha - g(u)\sin\alpha + A, f(u)\sin\alpha + g(u)\cos\alpha + B) = 0 \qquad (5.3)$$

or

$$F(f(u, v)\cos\alpha + g(u, v)\sin\alpha + A, -f(u, v)\sin\alpha + g(u, v)\cos\alpha + B) = 0. \quad (5.4)$$

In the case of (5.3) the base has the coordinate system $x0y$ and carries the family of v curves constructed according to $\phi(x, y, z) = 0$. The transparency with coordinate system $x'0y'$ carries the u scale given by

$$x' = f(u) \quad \text{and} \quad y' = g(u).$$

A and B are the coordinates of the origin $0'$ with respect to $x0y$. α is the angle between $0x$ and $0'x'$. This case is illustrated in Figure 5.5.

In the case of (5.4) the base carries the binomial system (u, v) constructed from $x = f(u, v)$ and $y = g(u, v)$. The transparency carries the unscaled curve L constructed from $F(x', y') = 0$. In this case the coordinates of $0'$ in the xy system are A', B' and α is again the angle between $0x$ and $0'x'$. This case is illustrated in Figure 5.6. The relationships between A and B and A' and B' are given by

$$A = -A'\cos\alpha - B'\sin\alpha,$$

$$\text{and} \quad B = A'\sin\alpha - B'\cos\alpha.$$

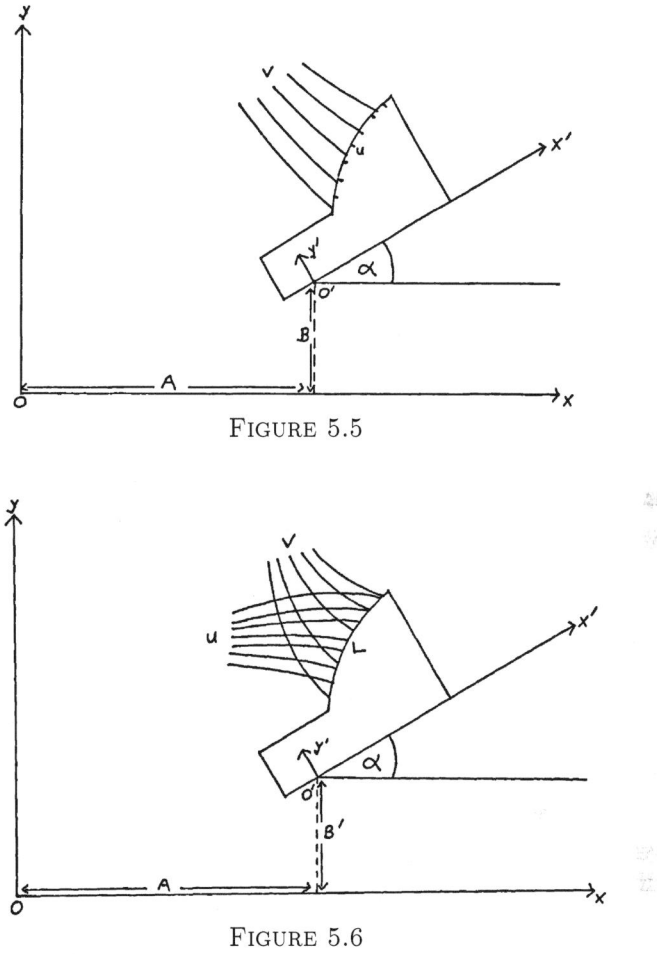

FIGURE 5.5

FIGURE 5.6

These two basic nomograms are used to tackle certain problems. Firstly interpolation: in the case of (5.3), (Figure 5.5), the transparency is arranged so that the points representing the values u_1, u_2, and u_3 which lie on it are in contact with the corresponding points v_1, v_2, and v_3 on the base. Then intermediate corresponding values of u and v may be read. In the case of (5.4),

(Figure 5.6), the method is similar except that the values of u and v both lie on the base, the unscaled curve L serving to join corresponding values.

The second problem which may be solved assumes that sets of corresponding values of u and v have been found experimentally and it is required to find the parameters A, B and α of (5.3) or (5.4). The technique is similar to that used for interpolation. In (5.3), (Figure 5.5), the transparency is positioned by eye so that $u_1, u_2, \ldots, u_n$ correspond respectively to $v_1, v_2, \ldots, v_n$. The position of the transparency then determines the parameters. For (5.4), (Figure 5.6), the correspondence is between (u_1, v_1) and L, (u_2, v_2) and L and so on up to (u_n, v_n) and L, giving A', B', and α. A and B can then be found from the relationships given earlier.

The third problem concerns the study of the errors involved when it is necessary to approximate the function $\phi(u)$ by a three parameter family $P(u, A, B, \alpha)$. This is done in terms of (5.4). For the absolute error we have

$$\Delta = \phi(u) - P(u, A, B, \alpha)$$

and for the relative error

$$\delta = \frac{\phi(u) - P(u, A, B, \alpha)}{\phi(u)}$$

and it is clearly desirable that the maximum values of $|\Delta|$ or $|\delta|$ should be a minimum.

Taking the case of absolute error as an example,

$$\Delta = \phi(u) - P(u, A, B, \alpha)$$

can be written as

$$\bar{P}(\phi(u) - \Delta, u, A, B, \alpha) = 0.$$

Replacing in (5.4) the variable v by $\phi(u) - \Delta$, a form which can be nomographed is obtained, namely

$$F\left(f(u, \phi(u) - \Delta)\cos\alpha + g(u, \phi(u) - \Delta)\sin\alpha + A,\right.$$
$$\left.-f(u, \phi(u) - \Delta)\sin\alpha + g(u, \phi(u) - \Delta)\cos\alpha + B\right) = 0$$

The binomial field on the base is given by,

$$x = f(u, \phi(u) - \Delta) \quad \text{and} \quad y = g(u, \phi(u) - \Delta).$$

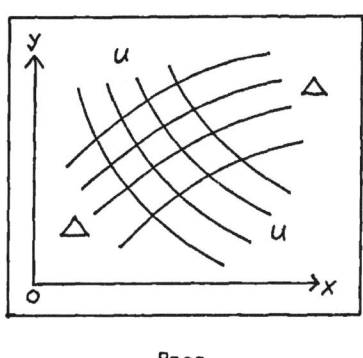
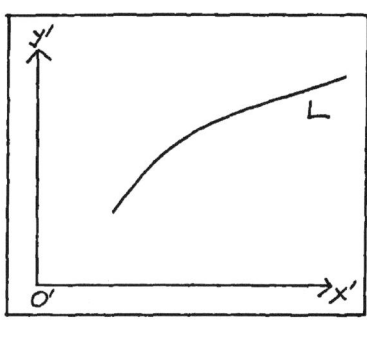

Base   **Transparency**

FIGURE 5.7

The unscaled line L on the transparency is again given by $F(x', y') = 0$. The form of the nomogram is illustrated in Figure 5.7.

Any position of the transparency on the base will show graphically the relationship between Δ and u. Displacing the transparency enables a graphical representation to be obtained showing the influence of A', B', and α on Δ. The argument in the case of relative error is similar.

The final problem concerns the approximation of a given function by a four parameter function; i.e., to find values of A, B, C, and α such that within given limits of u, the equation

$$\phi(u) = P(u, A, B, C\alpha)$$

becomes an identity in u. Writing the above equation in the form,

$$\overline{P}(u, \phi(u), A, B, C\alpha) = 0$$

and replacing, in (5.4), $f(u, v)$ by $f(u, \phi(u), C)$ and $g(u, v)$ by $g(u, \phi(u), C)$, we obtain,

$$F\left(f(u, \phi(u), C)\cos\alpha + g(u, \phi(u), C)\sin\alpha + A,\right.$$
$$\left.-f(u, \phi(u), C)\sin\alpha + g(u, \phi(u), C)\cos\alpha + B\right) = 0.$$

The base of the nomogram contains the field (u, C) given by

$$x = f(u, \phi(u), C) \quad \text{and} \quad y = g(u, \phi(u), C)$$

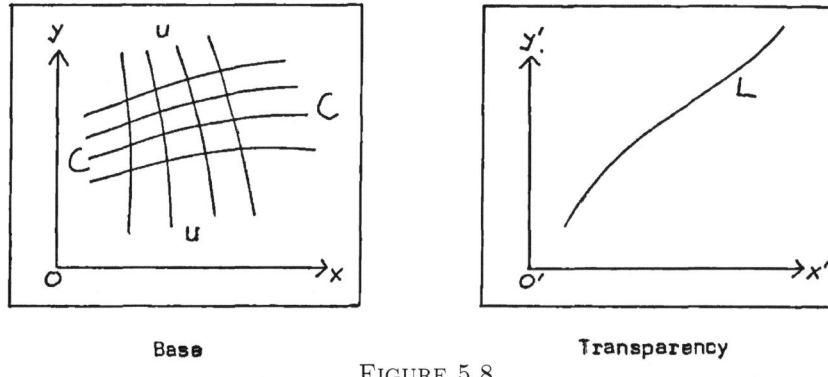

Base   Transparency

FIGURE 5.8

and the transparency will contain the usual unscaled line L, as shown in Figure 5.8.

If the sought for identity exists within certain limits then, within those limits, it will be possible to arrange the transparency so that part of L coincides with the appropriate part of a line C. More likely, an approximate coincidence will be the best that can be obtained; when this is obtained C is known and A, B and α can be determined. This problem can also be solved using (5.3). Khovanskii develops special cases in some detail using the principles already developed.

Still on the subject of approximate nomograms, Khovanskii considers the problem of constructing an approximate alignment nomogram for the two equations,

$$f_1 f_3 + f_2 g_3 + h_3 = 0 \qquad (5.5)$$
$$\text{and} \quad F_1 f_4 + F_2 g_4 + h_4 = 0 \qquad (5.6)$$

in which the scales of α_1 and α_2 are to be combined. If the relationship between f_1 and F_1 and that between f_2 and F_2 is linear so that,

$$F_1 = A_1 + B_1 f_1 \qquad (5.7)$$
$$\text{and} \quad F_2 = A_2 + B_2 f_2, \qquad (5.8)$$

then an accurate nomogram can be constructed. In this case (5.6) becomes,

$$f_1 B_1 f_4 + f_2 B_2 g_4 + (A_1 f_4 + A_2 g_4 + h_4) = 0$$

and a nomogram with combined scales in α_1 and α_2 can be constructed for this equation combined with (5.5).

The question arises of how to construct the nomogram with combined scales if (5.7) and (5.8) are only approximate. When scale factors are introduced, the alignment determinants for (5.5) and (5.6) may be written,

$$\begin{vmatrix} -H & m'(f_1 - a') & 1 \\ H & n'(f_2 - b') & 1 \\ \dfrac{H(g_3 m' - f_3 n')}{g_3 m' + f_3 n'} & \dfrac{-m'n'(h_3 + a'f_3 + b'g_3)}{g_3 m' + f_3 n'} & 1 \end{vmatrix} = 0$$

and

$$\begin{vmatrix} -H & m'(f_1 - a') & 1 \\ H & n'(f_2 - b') & 1 \\ \dfrac{H(g_4 m'' - f_4 n'')}{g_4 m'' + f_4 n''} & \dfrac{-m''n''(h_4 + a''f_4 + b''g_3)}{g_4 m'' + f_4 n''} & 1 \end{vmatrix} = 0.$$

Khovanskii replaces them by,

$$\begin{vmatrix} -H & \dfrac{m'(f_1 - a') + m''(F_1 - a'')}{2} & 1 \\ H & \dfrac{n'(f_2 - b') + n''(F_2 - b'')}{2} & 1 \\ \dfrac{H(g_3 m' - f_3 n')}{g_3 m' + f_3 n'} & \dfrac{-m'n'(h_3 + a'f_3 + b'g_3)}{g_3 m' + f_3 n'} & 1 \end{vmatrix} = 0 \qquad (5.9)$$

and

$$\begin{vmatrix} -H & \dfrac{m'(f_1 - a') + m''(F_1 - a'')}{2} & 1 \\ H & \dfrac{n'(f_2 - b') + n''(F_2 - b'')}{2} & 1 \\ \dfrac{H(g_4 m'' - f_4 n'')}{g_4 m'' + f_4 n''} & \dfrac{-m''n''(h_4 + a''f_4 + b''g_4)}{g_4 m'' + f_4 n''} & 1 \end{vmatrix} = 0 \qquad (5.10)$$

which give approximate scales for α_1 and α_2.

Consider the α_1 scale. There will be no error in the x direction since the scale is on a vertical straight line. To examine the error in the y direction let

$$y_1' = m'(f_1 - a'), \quad y_1'' = m''(F_1 - a''),$$

and

$$y_1 = \frac{m'(f_1 - a') + m''(F_1 - a'')}{2},$$

then

$$\Delta y_1' = y_1' - y_1 = \frac{m'(f_1 - a') - m''(F_1 - a'')}{2}$$

and

$$\Delta y_1'' = y_1'' - y_1 = \frac{m''(F_1 - a'') - m'(f_1 - a')}{2}.$$

Hence

$$|\Delta y_1'| = |\Delta y_1''| \quad \text{and} \quad \Delta y_1' = -\Delta y_1''.$$

If $\delta_1'' = \frac{\Delta y_1''}{m''}$, then $2\delta_1'' = F_1 - A_1 f_1 - B_1$ where

$$A_1 = \frac{m'}{m''} \quad \text{and} \quad B_1 = a'' - \frac{a'm'}{m''}.$$

Since A_1 and B_1 are arbitrary they can be chosen so that the maximum value of $|\delta_1''|$ is a minimum within given limits of α_1. The appropriate values of A_1 and B_1, denoted by A_1^* and B_1^*, may be found nomographically.

A similar argument applied to the α_2 scale gives

$$2\delta_2'' = F_2 - A_2 f_2 - B_2$$

where

$$\delta_2'' = \frac{\Delta y_2''}{n''}, \quad A_2 = \frac{n'}{n''}, \quad \text{and} \quad B_2 = b'' - \frac{n'b'}{n''}.$$

If m', a', n', and b' are eliminated from (5.9) and (5.10), by using the expressions for A_1, B_1, A_2, and B_2, the determinants take the form

$$\begin{vmatrix} -H & m(\phi_1 - a) & 1 \\ H & n(\phi_2 - b) & 1 \\ \dfrac{H(g_{34}m - f_{34}n)}{g_{34}m + f_{34}n} & \dfrac{-mn(h_{34} + af_{34} + bg_{34})}{g_{34}m + f_{34}n} & 1 \end{vmatrix} = 0 \qquad (5.11)$$

where

$$\phi_1 = \frac{A_1^* f_1 + B_1^* + F_1}{2} \quad \text{and} \quad \phi_2 = \frac{A_2^* f_2 + B_2^* + F_2}{2}$$

and f_{34}, g_{34}, and h_{34} have different values for the scale α_3 and α_4 as shown below.

Scale	f_{34}	g_{34}	h_{34}
α_3	$A_2^* f_3$	$A_1^* g_3$	$A_1^* A_2^* h_3 - A_2^* B_1^* f_3 - A_1^* B_2^* g_3$
α_4	f_4	g_4	h_4

TABLE 5.3.

It is the parameters A_1^*, B_1^*, A_2^*, and B_2^* which determine the quality of the approximation. As has been indicated these are found by a separate operation. When they have been found the nomogram may be constructed in the normal way.

In equation (5.11), if $m = n = H = 1$ and $a = b = 0$, then

$$\begin{vmatrix} -1 & \phi_1 & 1 \\ 1 & \phi_2 & 1 \\ \dfrac{g_{34} - f_{34}}{g_{34} + f_{34}} & \dfrac{-h_{34}}{g_{34} + f_{34}} & 1 \end{vmatrix} = 0$$

or

$$\phi_1 f_{34} + \phi_2 g_{34} + h_{34} = 0. \tag{5.12}$$

Therefore the construction of a combined scale nomogram for (5.5) and (5.6) is based on the approximate substitution for these equations of equation (5.12), appropriately interpreted.

Finally, some magnitude for the errors is required. Khovanskii suggests that,

$$\Delta y_1^* \leq \Delta y_{\text{error}} \quad \text{and} \quad \Delta y_2^* \leq \Delta y_{\text{error}}$$

where $\Delta y_{\text{error}} = 0.2$mm and $*$ is used as before; i.e., Δy_1^* the maximum value of $|\Delta y_1|$. This gives

$$m \leq \frac{\Delta y_{\text{error}}}{\delta_1^*} \quad \text{and} \quad n \leq \frac{\Delta y_{\text{error}}}{\delta_2^*}.$$

The final section of Khovanskii's paper considers the use of nomograms for the investigation of functional relations.

Such nomograms require different attributes from those constructed for calculation. For example, the limits of the variables must be complete in order to cover all possible cases while preference must be given to the type of nomogram which shows most completely the interesting characteristics of the relationships being investigated. Alignment nomograms lose their prime position in the latter respect. Intersection nomograms and those with an oriented transparency are often more convenient.

Some types of investigations which can be carried out with nomograms are,

    a. To examine the influence of one parameter on the others.
    b. To give a geometric illustration of some important, already well known, property of a formula.
    c. To find hitherto unknown properties of a given relationship.

Khovanskii gives a lengthy illustration of this aspect of nomograms drawn from the flow of fluid in channels of different cross-sections. Hydraulic calculations are a particular interest of Khovanskii. An extract will illustrate the type of enlightenment that can be shed.

For the case of a round channel the transparency has the form shown in Figure 5.9. Here η is a dimensionless parameter, V is the average velocity and i is the gradient.

Although the base of the nomogram contains many lines representing the different variables of the problem, it will suffice to consider only the two variables d and I. The diameter d is represented by parallel straight lines which are also parallel to the transparency scales of V and i. I is represented by a straight line perpendicular to V, the purpose of which is to indicate the value

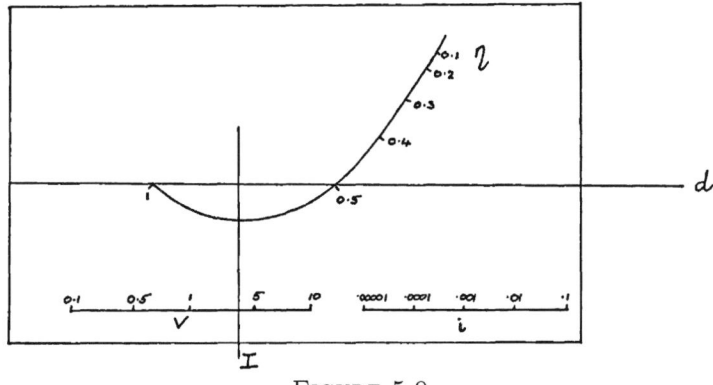

FIGURE 5.9

of V. Some of the lines d will intersect the η curve in two places, in particular one of the lines intersects it at 1 and 0.5. The interpretation of this is that the velocity in a pipe of a given diameter is the same at full loading as at half loading, a result which may not be so readily seen by any other analysis.

A most interesting article by T. Steyskalova takes some results from W. Blaschke's book on the theory of nets [1] [8] and demonstrates their application to nomography [129]. The application is elegant but to appreciate this elegance it is necessary to understand something of Blaschke's theory.

We suppose in a region G of the xy plane three families of curves σ_1, σ_2, and σ_3 given by $u_i(x,y) =$ constant, where u_i is analytic in G. The system has the following properties:

    i. $\left(\frac{\partial u_i}{\partial x}\right)^2 + \left(\frac{\partial u_i}{\partial y}\right)^2 \neq 0$ at every point of G

    ii. through every point of G passes one, and only one, curve of each family,

    iii. the Jacobian $\frac{\partial(u_j, u_k)}{\partial(x,y)} \neq 0$ with $j \neq k$,

---

[1]The German word GEWEBE has as its English counterpart the word webbing. The Russians use the word CETEN which is translated as nets. I use the word *net* and hope that no confusion will arise

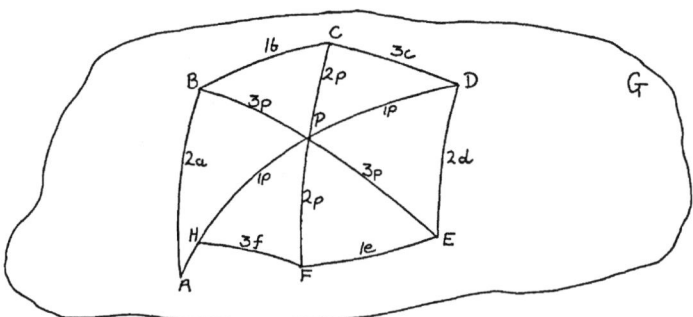

FIGURE 5.10. Brianchon's Diagram

   iv. any two curves of different families have not more then one common
       point,
   v. within G every curve is continuous.

Such a system is known as a *triple system*.

   The simplest triple system is the regular net which is a system consisting
of three families of parallel straight lines with the lines of different families
meeting at $60°$. Obviously such a net forms regular hexagons whose sides and
diagonals are straight lines of the regular net. Steyskalova calls hexagons of
this type diagrams B, after Brianchon. Diagrams B are not necessarily regular
hexagons. In fact, diagrams B are those figures which are constructed in the
following way.

   Given a triple system in a region G. Take a point P inside G and draw a
line through it for each family, denoting them by $1p$, $2p$, and $3p$, where 1, 2,
and 3 refer to the family to which the line belongs. On line $1p$ select the point
A and draw the line $2a$ to intersect $3p$ at B. Then draw line $1b$ to intersect $2p$
at C. In a similar way obtain the points $D(3c\&1p)$, $E(2d\&3p)$, $F(1e\&2p)$, and
$H(3f\&1p)$ (Figure 5.10).

   The point H may coincide with A, thus closing the diagram, but this is
not necessary. Triple systems in which all diagrams B are closed are called
hexagonic.

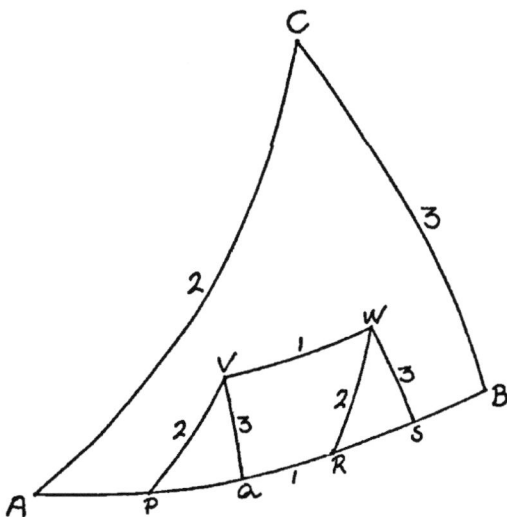

FIGURE 5.11. Steyskalova's Coordinate Triangle

A curvilinear triangle ABC having sides which are curves of the first, second and third families is called a coordinate triangle. Points P, Q, R, and S are chosen on the side AB, which belongs to the first family. Through P and R lines of the second family are drawn and through Q and S lines of the third family are drawn, giving points of intersection V and W. If V and W lie on the same line of family 1; i.e., $1v = 1w$, then the arcs PQ and RS are said to be equal. The property of equality is symmetric and transitive (Figure 5.11). If the curve $1v$ lies between the curves $1w$ and AB then PQ is said to be smaller than RS.

In the case of the regular net the coordinate triangle is an equilateral triangle and the definition of equality coincides with the ordinary concept of equality. Taking this coordinate triangle and dividing its sides into integer n parts and then drawing lines of all families through the points of division, a diagram of the type shown in Figure 5.12 is obtained. Steyskalova calls this diagram D as he also does all topological forms of it.

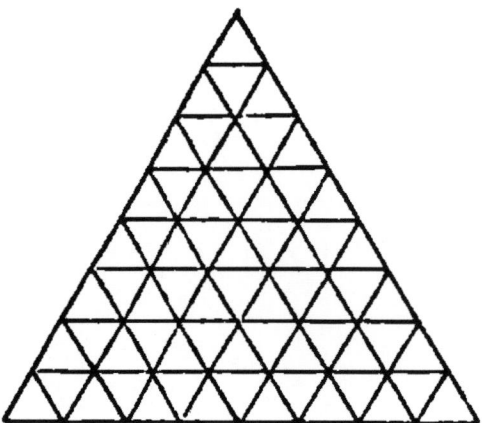

FIGURE 5.12. Steyskalova's Diagram D_0 with $n = 8$

He gives two lemmas.

    I. Diagrams D can be constructed from any hexagonic net.
    II. In hexagonic nets the definition of equality of curvilinear segments possesses the property of addition.

He then proves a basic theorem for hexagonic nets.

    THEOREM. A given triple system can be nomologically mapped into a regular net when, and only when, all diagrams B are closed; i.e., when they are hexagonic.

    In addition to diagram B, two other diagrams are considered, Thomson's, called diagram T, and Rademayster's, called diagram R. The three diagrams are illustrated in Figure 5.13. Straight lines are used for simplicity.

    A second theorem is proved for diagrams T and R.

    THEOREM. For a triple system to be a hexagonic net it is necessary and sufficient for every diagram T (or R) to be closed.

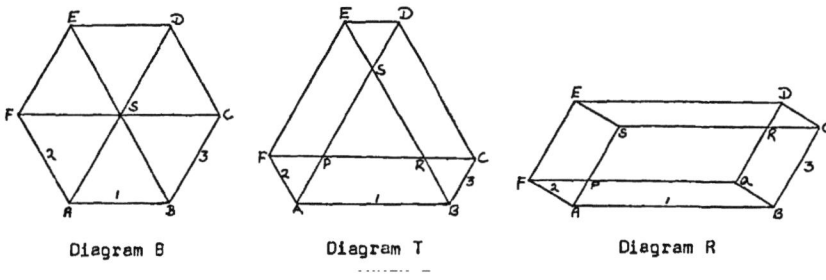

Diagram B   Diagram T   Diagram R

FIGURE 5.13

Features of the proof which are worth repeating here are that,

i. Diagram B is a particular case of diagram T in which P, R, and S coincide.

ii. Diagram B is a particular case of diagram R in which PQR and S coincide.

It is seen that all triple systems in which all diagrams T (or R or B) are closed can be mapped topologically onto a regular net. This is of significance to nomography. An intersection nomogram having curves which intersect in vary acute angles can lead to substantial errors in the results. Then it is desirable to increase the size of these small angles. An angle of 60° is ideal and fortunately this is the size of the angle contained in a regular net. It is therefore necessary to be able to recognize those equations which can be represented as a regular net. Such equations are equations of the third nomographic order. This leads to the third theorem of the paper,

THEOREM. For an equation $F(x, y, z) = 0$ to be of the third nomographic order it is necessary and sufficient that the following six equalities are such that each is a consequence of the other five,

$$F(x_0, y_1, z_1) = 0, \qquad F(x_0, y_2, z_2) = 0, \qquad F(x_1, y_2, z_3) = 0, \qquad (5.13)$$
$$F(x_2, y_1, z_3) = 0, \qquad F(x_2, y_0, z_2) = 0, \qquad F(x_1, y_0, z_1) = 0.$$

To see what this theorem is stating, assume an intersection nomogram to be constructed for $F(x, y, z) = 0$. Take three of the z lines z_1, z_2 and z_3, then

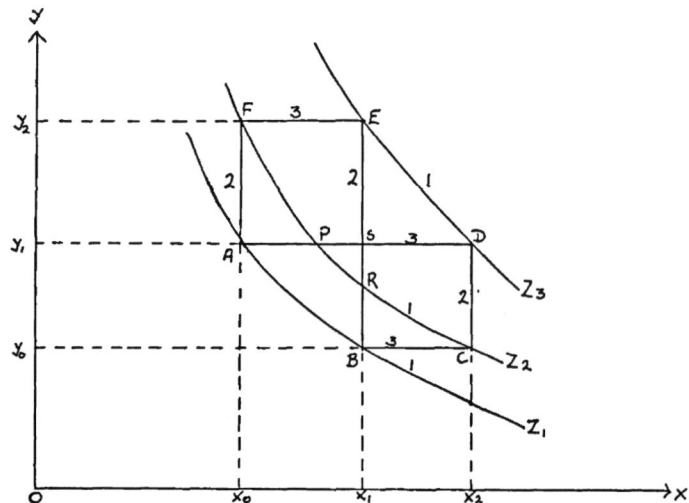

FIGURE 5.14. Thomson's diagram on an intersection nomogram for an equation of the third nomographic order

each of the equations (5.13) of the form $F(x_i, y_j, z_k) = 0$ indicates that the lines
$x = x_i$, $y = y_j$, and $z = z_k$ intersect in one point, or, that Thomson's diagram
is closed, as illustrated in Figure 5.14. If all of the Thomson diagrams are
closed, we know that the nomogram forms a hexagonic net and can therefore
be transformed into a regular net.

It is of value to examine the proof of this theorem. For sufficiency it is
assumed that $F(x, y, z) = 0$ leads to a nomogram on a hexagonic net and it is
then necessary to show that $F(x, y, z) = 0$ is of the third nomographic order.
In proving the basic theorem of hexagonic nets the result had been obtained
that for each line of the third family the relation $X + Y = $ constant held, where
X and Y were the values of the lines of the other two families intersecting on
the line of the third family at a particular point. Letting that constant be Z
for the corresponding line of the third family, the relation becomes $X + Y = Z$
which is of the third nomographic order.

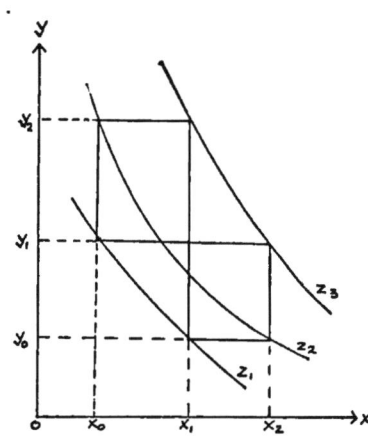

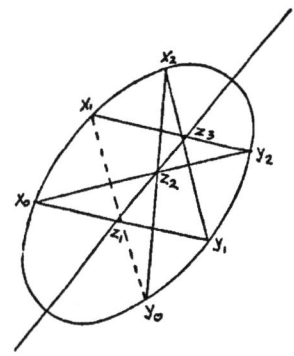

Thomson's diagram on third
order intersection nomogram.

Conical alignment nomogram
corresponding to Thomson's
diagram.

......

FIGURE 5.15

It is the necessary part of the proof which is so illuminating for nomography. Suppose that an equation of the third order is given. It is then necessary to show that each of the equalities (5.13) is a consequence of the other five.

Clark has shown that a third order equation can also be represented by a nomogram having a conic section and straight line (Chapter 3, Section 4). The conical alignment nomogram is the dual of the third order intersection nomogram, both of which are illustrated in Figure 5.15.

On the intersection nomogram $F(x_i, y_j, z_k) = 0$ indicates that the lines $x = x_i$, $y = y_j$, and $z = z_k$ pass through one point. On the alignment nomogram the three points x_i, y_j, and z_k lie on a straight line. Assuming that the first five equalities of (5.13) hold, it is necessary to prove the sixth. Suppose that the points x_0, x_1, x_2, y_0, y_1, y_2, z_1, z_2, and z_3 satisfy the first five. It is necessary to show that on the alignment chart the line joining x_1 and y_0 also passes through z_1; i.e., that $F(x_1, y_0, z_1) = 0$. But this is so by virtue of Pascal's theorem

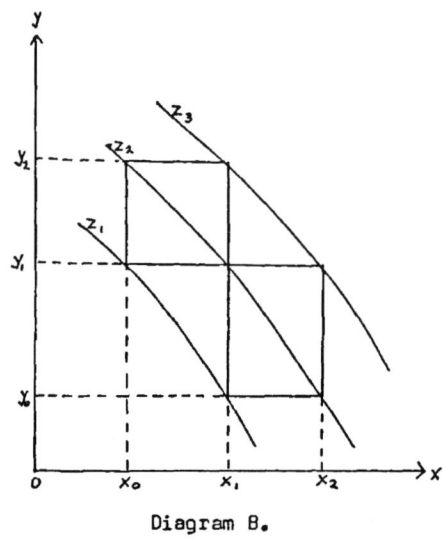

Diagram B.

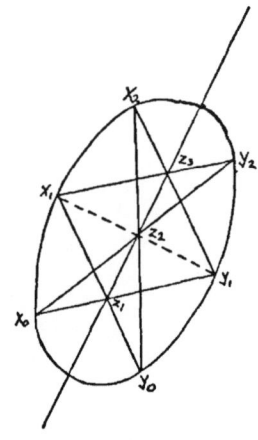

Alignment nomogram
corresponding to
diagram B.

FIGURE 5.16

of projective geometry, the points z_1, z_2, and z_3 lying on Pascal's line. This completes the proof.

Since Thomson's diagram corresponds to Pascal's configuration it is natural to enquire whether Brianchon's and Rademayster's do also.

Diagram B corresponds to the particular case when x_1, y_1 and z_2 are on a straight line as shown in Figure 5.16.

Steyskalova attributes to G.E. Džems-Levi the following result for diagram R.

On some curve K of the second degree cut by the straight line p draw the quadrilateral x_1, y_1, x_2, y_2 as in Figure 5.17. Let the line p intersect the sides of the quadrilateral at z_1, z_2, z_3

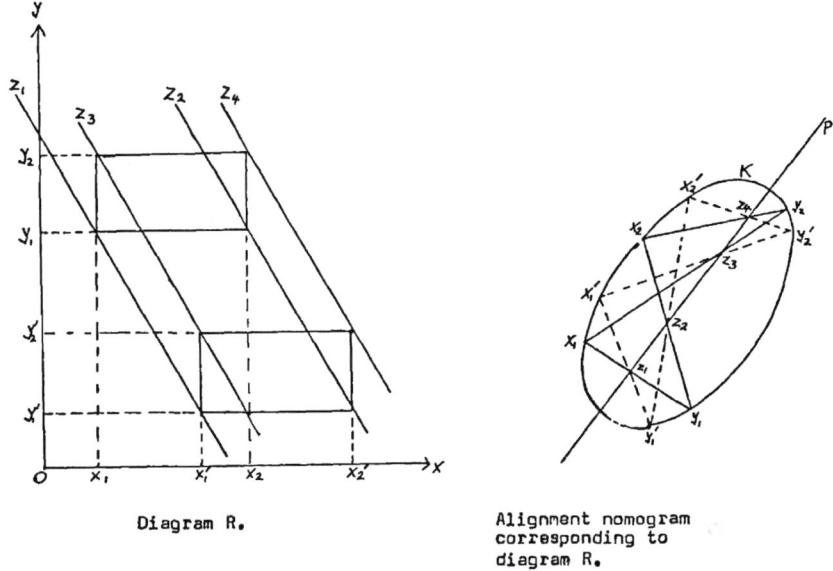

Diagram R.

Alignment nomogram
corresponding to
diagram R.

FIGURE 5.17

and z_4. Then every quadrilateral C with sides passing through
the points z_1, z_2, z_3 and z_4 of p and vortices on K will close if
just one such quadrilateral C_0 will close.

The proof follows immediately from the closure of diagram R since each quadrilateral

$$C((x_1, y_1), (x_1, y_2), (x_2, y_2), (x_2, y_1))$$

and

$$C_0((x_1', y_1'), (x_1', y_2'), (x_2', y_2'), (x_2', y_1'))$$

corresponds to the hexagonic net of diagram R.

5.2. Theoretical Considerations

Wilhelm Blaschke, whose theory of nets had proved to be of such interest to Steyskalova, published a very short paper on nomography in 1956 [7]. Blaschke addressed himself to some fundamental problems and viewed them in the light of his theory of nets. Given three families of curves $u_j(x, y)=$ constant he considers their nomogram given by the relation $T(u_1, u_2, u_3) = 0$. His first problem is to find the best nomogram, by which he means the net best suited for its representation. The sought for solution is, of course, a net consisting of straight lines. The second problem concerns uniqueness and asks the question whether two rectilinear nets representing the same equation are necessarily equivalent. His final question is to ask whether a nomogram is the optimum nomogram in the sense that some measure of the value of the nomogram is optimal.

Blaschke does not solve these problems in the strictest sense but illuminates them. On the question of producing a rectilinear net he uses the differential calculus to investigate invariants, obtained from the curvature, of topologically equivalent diagrams and reaches the conclusion that only at the ninth derivative of T are we able to expect the conditions for the rectification of a net, a rather sobering thought. For optimality, he looks to the calculus of variations. His solution is to minimize the integral,

$$J = \int \left(K_1^2 + K_2^2 + K_3^2 \right) \, dA$$

in which dA is the element of the surface at a point (x, y) and K_1, K_2, and K_3 are the curvatures of the curves of the net at the point (x, y).

There is no doubt that Blaschke's paper is important for the theory of nomography. From the standpoint of this thesis it is important in that once again the difficulty of the problem of anamorphosis is brought out.

Evidence that some of the theoretical problems raised by nomography are still very much alive is contained in a paper which appeared as recently as 1976 [14]. In this paper R.C. Buck discusses approximate complexity and functional representation. The section of particular interest to nomography addresses itself to the question of whether a given function $F(x, y)$ can be expressed in terms of these functions, each of one variable; i.e., can we write,

$$F(x, y) = f(\phi(x) + \psi(y))? \tag{5.14}$$

If (5.14) is a valid expression then it follows that $z = F(x, y)$ can be expressed as

$$z = f(\phi(x) + \psi(y))$$

with obvious nomographic advantages.

Buck's treatment is rigorous and proceeds through analysis. Some of his results are exactly those given by the Russian, T. Steyskalova, seventeen years earlier [**129**]. Steyskalova's results were obtained from Blaschke's theory of nets but Buck makes no mention of either Steyskalova or Blaschke and it seems probable that he was unaware of their work.

What follows is a brief outline of Buck's results from a nomographic viewpoint. It does not purport to do justice to the analytical rigor of his paper since to do that would be a diversion from the main theme of this thesis, but we can note that Buck's treatment once again highlights the difficult theoretical problems associated with nomography.

The first result is obvious enough. Buck shows that if $F(x, y)$ has the form (5.14) then it must satisfy the differential equation,

$$(F_x F_y)(f_x F_{xyy} - F_y F_{xxy}) + F_{xy}(F_y^2 F_{xx} - f_x^2 F_{yy}) = 0. \tag{5.15}$$

Much of Buck's analysis is concerned with two classes of functions F_0 and F_w. Both classes are of functions with the format given by (5.14) but F are those representable with continuous f, ϕ, and ψ, while F_w refers to those in which ϕ and ψ are continuous but f is unrestricted. He expresses the problem in the form of a mapping diagram which is worth repeating for its simplicity. He asks whether there are functions h and f such that Figure 5.18 commutes.

The function h belongs to the class of continuous functions of the form

$$h(x, y) = \phi(x) + \psi(y)$$

and f is unrestricted.

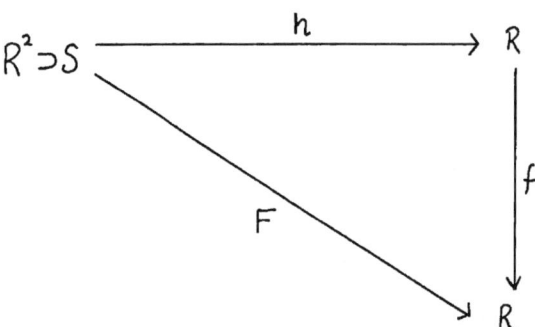

FIGURE 5.18

Buck sets himself the task of looking for properties that distinguish members of F, from other functions. The mapping of Figure 5.18 indicates that such properties must arise from the form of h. If functions $h(x, y) = \phi(x) + \psi(y)$ are defined on a rectangle S and P_1, P_2, P_3, and P_4 are successive vertices of a rectangle in S, with edges vertical and horizontal, then he observes that

$$h(P_1) - h(P_2) + h(P_3) - h(P_4) = 0 \tag{5.16}$$

and also that the converse holds. The property can be extended to any chain of $2n$ points P_i which are vertices of a closed polygon in S with edges that are successively vertical and horizontal. The case for $n = 4$ is illustrated in Figure 5.18.

The functions h defined on S form a proper closed subspace Y. Buck gives the following theorem:

THEOREM. If $F \in F_w(S)$ then thin connected sublevel sets for F in S must be Y sets.

This leads to an interesting result. Suppose that

$$h(x, y) = \phi(x) + \psi(y)$$

is constant on a vertical segment α in S. Then $\psi(y)$ must be constant on the vertical segment and h will then be constant on every segment parallel to α. The above theorem then suggests that if F is constant on a vertical segment,

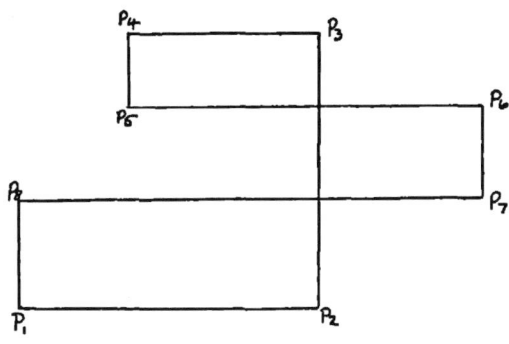

FIGURE 5.19

then it must be constant on every parallel segment. Now consider

$$F(x, y) = (x - c)^2 e^y$$

where c is constant. It is easy to show that it satisfies the condition (5.15). However, consider the value of $F(x, y)$ at the four points $(0, 0)$, $(c, 0)$, $(c, 1)$, and $(0, 1)$ and at any point on the side $x = c$ of the rectangle formed by these points. It is obvious that $F(x, y) = 0$ on the side $x = c$ but, on the side $x = 0$, $F(x, y)$ varies from c^2 at $y = 0$ to $c^2 e$ at $y = 1$. Thus, F cannot belong to the class $F_w(S)$ on any open rectangle S that contains the line $x = c$. This counter-example shows that, although it is necessary that (5.14) should satisfy the differential equation (5.15), the condition is not sufficient.

These, and other considerations, lead Buck to seek a local property for the Y sets. He produces what he calls *the six point construction*. This is based on an extension to six points of the idea given by (5.16); i.e.,

$$h(P_1) - h(P_2) + h(P_3) - h(P_4) + h(P_5) - h(P_6) = 0$$

located as in Figure 5.20.

Now, if $h(P_1) = h(P_4)$ and $h(P_3) = h(P_6)$ then it must be true that $h(P_2) = h(P_5)$. It therefore follows that, given two points, one on each of two level lines, a geometric construction will produce a pair of points lying on a third level line.

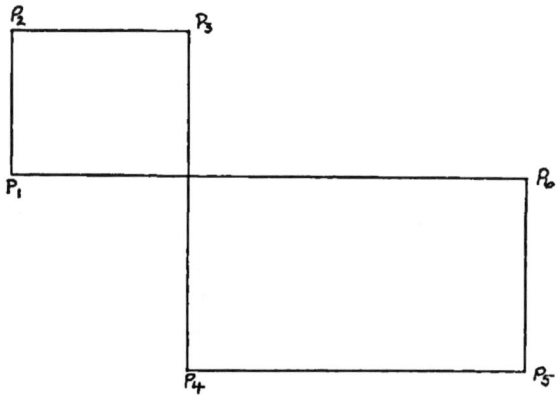

FIGURE 5.20

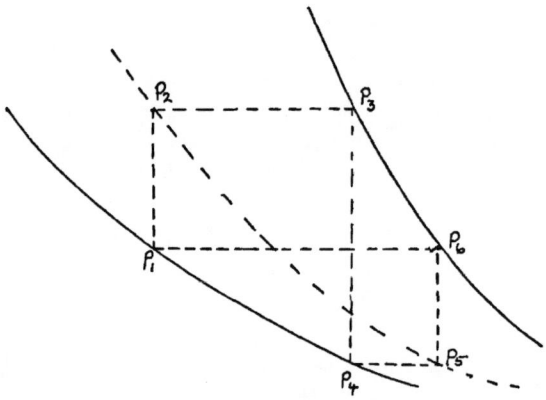

FIGURE 5.21

Suppose that P_1 and P_3 are given points on two level lines, (Figure 5.21). If vertical and horizontal lines are drawn through these points, as shown by the dotted lines, two further points on the given level lines will be obtained, P_4 and P_6, and two points on a third level line, P_2 and P_5.

This is the first of Buck's results which had previously been obtained by Steyskalova. The diagram is nothing more than Thomson's diagram on an intersection nomogram for an equation of the third nomographic order (Figure 5.14).

Buck does take the matter further, however. He proves a theorem which states that if $h(x, y) = \phi(x) + \psi(y)$ on the rectangle S, where $\phi(x)$ and $\psi(y)$ are continuous and strictly increasing, then the six point construction applies locally everywhere in S.

A further theorem states that if $F \in F_w(S)$, where $S = I \times J$, and F is separately univalent on S; i.e., separately one to one mappings, then for any a and c in I and b and d in J, with $|a - c|$ and $|b - d|$ sufficiently small, there must exist x and y near a and b respectively, such that,

$$F(a, b) = F(c, y),$$
$$F(c, d) = F(x, b), \tag{5.17}$$
$$\text{and} \quad F(a, d) = F(x, y)$$

This result is a statement of the same property expressed by the third theorem of Steyskalova's paper [**129**]. Although in a different form, I hold that equations (5.17) are equivalent to equations (5.13) and I outline my reason for this in Appendix D. However, once again Buck takes the matter further and illustrates how the results may be applied.

He considers the function $F(x, y) = x^2 + xy + y^2$ and shows that it is not locally nomographic anywhere in the first quadrant.

The method is to take $a > 0$, $b > 0$, and c and d such that $a < c$ and $b < d$. Equations (5.17) yield,

$$c^2 + cy + y^2 = a^2 + ab + b^2,$$
$$x^2 + bx + b^2 = c^2 + cd + d^2, \tag{5.18}$$
$$\text{and} \quad x^2 + xy + y^2 = a^2 + ad + d^2$$

and it is required to show that there exist infinitely many c and d such that the system (5.18) is inconsistent. Clearly in the general case this can be very tedious. He is led to the following conjecture,

"A polynomial $F(x,y)$ will not belong to the class F_w on any open set unless it satisfies the differential equation (5.15) and can be written as $f(\phi(x)+\psi(y))$ with f, ϕ, and ψ polynomials."

Buck investigates whether a specific function G can be approximated uniformly on compact sets by nomographic functions of the set F_w.

He arrives at a criterion for approximate representation which is expressed in his Theorem 15.

THEOREM. Let G be continuous on $S = I \times J$, where $I = [a, b]$ and $J = [c, d]$. Suppose also that $G_x > \sigma$ and $G_y > \sigma$. Suppose also that $G_x > \sigma$ and $G_y > \sigma$ on S where $\sigma > 0$. Let (u, v) be any point on S such that $|2u - (a + b)| < 2L/3$ and $|2v - (c + d)| < 2L/3$ where L is the length of the shorter side of S. If G lies in the uniform closure of $F_w(S)$ then for any sufficiently small ε, such as $\varepsilon < L\sigma/12$, one of the following statements must hold:

(i) there exist x and y in J such that
$$|G(a, x) - G(u, c)| < \varepsilon,$$
$$|G(a, y) - G(b, c)| < \varepsilon,$$
$$\text{and} \quad |G(b, x) - G(u, y)| < \varepsilon$$

or (ii) there exist x and y in I such that
$$|G(x, c) - G(a, v)| < \varepsilon,$$
$$|G(y, c) - G(a, d)| < \varepsilon,$$
$$\text{and} \quad |G(x, d) - G(y, v)| < \varepsilon.$$

Buck demonstrates the effectiveness of this theorem by reference to the function
$$G(x, y) = x^2 + xy + y^2$$
as follows.

S is the unit square of Figure 5.22 and (u, v) is the point $(1.5, 0.5)$. The second statement of the theorem then says that the following system should

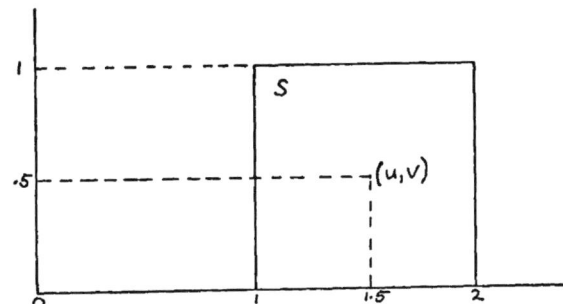

FIGURE 5.22. I: $a = 1$, $b = 2$ and J: $c = 0$, $d = 1$

have a solution for all sufficiently small ε:

$$|x^2 - 1.75| < \varepsilon$$
$$|y^2 - 3| < \varepsilon \tag{5.19}$$
$$|(x^2 + x + 1) - (y^2 + 0.5y + 0.25)| < \varepsilon.$$

A suitable value for σ in S is 0.5, $L = 1$ and, using the suggestion in the theorem, $\varepsilon < 0.5/12$, so we can take $\varepsilon = 0.04$. Two convenient values in I would be $x = \sqrt{1.75}$ and $y = \sqrt{3}$.

The first two equations of (5.19) are satisfied. The left hand side of the third equation becomes $|4.073 - 4.1161| = .043$ and therefore this equation does not hold. The equations (5.19) cannot have a simultaneous solution for small ε and therefore $x^2 + xy + y^2$ cannot be approximated uniformly on S by functions in the class F_w. By the use of small rectangles this result can be extended to cover the whole of the first quadrant.

Buck has performed an important service to theoretical nomography. Taken in conjunction with the parallel ideas of Steyskalova, Buck's paper suggests a way forward for further theoretical development.

5.3. Some Recent Nomograms

With the availability of electronic computing aids it may be thought that nomograms are no longer used. This, however, is not the case. The following three examples will show that there is still a place for nomograms. In addition, I have learned that quite recently nomograms have been used in the petroleum industry, to compute viscosity, and also by the meteorological office.

My first example comes from the Directorate of Overseas Surveys. This directorate has prepared a booklet devoted to nomograms for survey computations [28]. Fourteen nomograms of the alignment type are given, all for calculations frequently encountered by land surveyors; for example, the first three give the meteorological correction for the Tellurometer, Geodimeter and the Wild DI 10 Distormat respectively. An interesting feature is that the booklet is supplied with a nomogram reader, an ingenious device necessary to read one of the nomograms. The nomogram reader was developed by Mr. Bowring of the computing section of the directorate. The preface contains the following paragraph,

> "Nomograms are simple devices which, in the right circumstances, aid rapid computation with little loss of accuracy. Provided their accuracy limits are recognized, they can be used to solve many survey problems."

The second example is a set of nomograms for morphometric gravel analysis which appeared in 1977 [47]. These are the work of J.L. Van Genderen. The author points out that most methods of quantitative morphometric gravel analysis are very time consuming and as an illustration of this takes the computations necessary for the index of roundness when carried out in the field. He states that for one sample of 100 rock fragments it can take between one and one and a half hours. Even, he states, if done in the office using an electronic calculator it takes from one half to one hour to arrive at the required histogram. However, he claims, using one of his nomograms the raw data can be converted into the final histogram in less than ten minutes. The nomograms themselves are quite simple as can be seen from the formulae which they represent. They are for the index of roundness $I_r = 2r/L$; the index of flatness $I_f = (L+1)/2E$;

and the index of dissymmetry $I_d = AC/L$; in which r, L, E, and AC are all distances.

Before leaving Van Genderen's nomograms it is worth noting his comment on the advantage of a nomogram over other methods. He states that "the need is normally for a rapid method which can be employed on the spot, so that any anomalies or significant results can immediately be examined in the field." Although the situations are different, this view is very similar to that expressed by Capt. C.E.P. Sankey in 1911 on the advantages of nomograms on military service, ([123] and Chapter 3, Section 3).

The last example concerns the calculation of interest rates, a process which is often more complex than is generally recognized. The particular problem is the relationship between flat, nominal and effective rates of interest for which the formulae are,

$$e = (1 + i/n)^n - 1 \qquad\qquad (5.20)$$

$$f = \frac{1 + (iy - 1)(1 + i/n)^{ny}}{y((1 + i/n)^{ny} - 1)} \qquad\qquad (5.21)$$

where f is the flat rate, i the nominal rate, e the effective rate, y the number of years and n the number of installments per annum.

The problem lies in the fact that, although equation (5.20) can be rewritten to give i in terms of e, equation (5.21) cannot be similarly rearranged so that i is given explicitly in terms of f. Given f, i (or e) can only be determined iteratively. Such an iterative procedure had been given in the Bulletin of the Institute of Mathematics in 1979 and this appears to have prompted J. Rickard, of the University of Melbourne, to produce a nomogram for the same problem. His nomogram appeared in the October 1980 issue of the same bulletin [120]. It is elegant since it contains only straight lines, or very good approximations to straight lines, and is read along lines which are parallel to the axes. The accuracy requirements in such problems make nomograms quite suitable as a method of calculation. This last example illustrates that there are still areas of computation in which a nomogram is still the best choice.

CHAPTER 6

Conclusions

The question which springs to mind at the end of this thesis is "Has nomography a future?" In the widest sense I think it has not. It is true that in special cases, such as the three mentioned at the end of the last chapter, nomograms may still be the most suitable method of calculation, but as a general method of calculation their fate has been sealed by the advent of the pocket electronic calculator. This is not to say that whenever a calculator is used it will produce a result which in some way is better than that which would have been produced by a nomogram, but rather that the former has some kind of psychological advantage over the latter. An answer is displayed in lights and therefore has some veracity although the calculation which led to it may not have been at all appropriate. There is another reason for the demise of the nomogram; it is the difficulties associated with anamorphosis. Although the approximation techniques of Džems-Levi and other Russian workers might have improved the outlook had they appeared earlier, the truth is that they were too late and when they came the electronic computer and calculator were more attractive alternatives.

What of the application of modern technology to nomography? Two ideas seem to be worth some investigation. The first idea concerns the use of the graphical display techniques of modern microprocessors. Although I have suggested that computers heralded the end of nomography, I was then thinking of the large machines of the 1960's and 1970's. Using the high resolution display of the modern microprocessor, the construction of electronic nomograms would seem to be a possibility in those cases where very high accuracy is not required. Advantages of such nomograms would include programmable transformation and magnification, something that earlier workers never dreamed of.

The second idea is more fanciful. It is to simulate numerically within a computer some of the techniques of nomography. I see this as a way of getting reasonable approximations which would then be subject to refinement by one of the many techniques which exist for doing this. The advantage of such a method, if there is an advantage in it, would depend entirely upon how good the approximation was and how fast it could be arrived at.

As for work on the history of the subject, several avenues have been opened up by this thesis. A detailed study of geometric computation before 1840 would be a worthwhile undertaking. Then, there is the question of the use made of nomograms in the various disciplines. This has only been touched on here because it was peripheral to the main theme. Also, there are the characters themselves. A study of them would widen the field beyond nomography but it would be interesting to know something of the life and mathematical activities of Dr. J. Clark, a man of obvious mathematical skills. The same could be said of Grönwall, O.D. Kellogg, Warmus and Džems-Levi. Finally, of course there is still Hilbert's thirteenth problem.

APPENDIX A

Massau's and Lecornu's Conditions

A.1. Massau's Conditions

The following are Massau's conditions for $F(x, y, z) = 1$ to be expressed in the form

$$Z_1(z)X(x) + Z_2(z)Y(y) = 1.$$

Start with

$$Z_1 X + Z_2 Y = 1. \tag{A.1}$$

Differentiate (A.1) partially with respect to x,

$$Z_1 X' + (X Z_1' + Z_2' Y)p = 0. \tag{A.2}$$

Differentiate (A.1) partially with respect to y,

$$Z_2 Y' + (X Z_1' + Z_2' Y)q = 0. \tag{A.3}$$

Eliminate $X Z_1' + Z_2' Y$ from (A.2) and (A.3)

$$\frac{Z_1}{Z_2} = \frac{pY'}{qX'}. \tag{A.4}$$

This shows that z is a function of $\frac{pY'}{qX'}$ from which it follows that the Jacobian of z and $\frac{pY'}{qX'}$ is zero, in other words

$$\begin{vmatrix} p & \frac{\partial}{\partial x}\left(\frac{pY'}{qX'}\right) \\ q & \frac{\partial}{\partial y}\left(\frac{pY'}{qX'}\right) \end{vmatrix} = 0.$$

On expansion this gives

$$p\frac{Y''}{Y'} + q\frac{X''}{X'} = r\frac{q}{p} - 2s + t\frac{p}{q}$$

or

$$p\frac{Y''}{Y'} + q\frac{X''}{X'} = R. \tag{A.5}$$

Let $\frac{X''}{X'} = U_1$ and $\frac{Y''}{Y'} = U_2$. Then (A.5) becomes

$$pU_2 + qU_1 = R. \tag{A.6}$$

Differentiate (A.6) with respect to x

$$rU_2 + sU_1 + qU_1' = \frac{\partial R}{\partial x}. \tag{A.7}$$

Differentiate (A.6) with respect to y

$$sU_2 + pU_2' + tU_1 = \frac{\partial R}{\partial y}. \tag{A.8}$$

Differentiate (A.6) with respect to y

$$\frac{\partial r}{\partial y}U_2 + rU_2' + \frac{\partial s}{\partial y}U_1 + \frac{\partial q}{\partial y}U_1' = \frac{\partial^2 R}{\partial x \partial y}. \tag{A.9}$$

Equations (A.6), (A.7), (A.8) and (A.9) will give U_1, U_2, U_1', and U_2'.

$$U_1 = \frac{X''}{X'} \qquad \text{hence} \qquad \frac{\partial U_1}{\partial y} = 0$$

$$\text{and} \quad U_2 = \frac{Y''}{Y'} \qquad \text{hence} \qquad \frac{\partial U_2}{\partial y} = 0.$$

With these conditions satisfied, X and Y can be found. Z_1 and Z_2 are then found from (A.4) and (A.1).

A.2. Lecornu's Conditions

The following are Lecornu's conditions for

$$F(x, y, z) = 1$$

to be expressed in the form

$$Z_1(z)X(x) + Z_2(z)Y(y) = 1.$$

The first part of Lecornu's reasoning is the same as Massau's; from

$$Z_1X + Z_2Y = 1$$

he obtains (A.4) but in the form

$$\frac{q}{p} = \frac{Z_2 Y'}{Z_1 X'} \tag{A.10}$$

from which

$$\ln \frac{q}{p} = \ln \frac{Z_2}{Z_1} + \ln Y' - \ln X'.$$

The substitutions

$$T = \ln \frac{Z_2}{Z_1}, \quad f = -\ln X', \quad \text{and} \quad g = \ln Y'$$

are made giving

$$\ln \frac{q}{p} = T + g + f. \tag{A.11}$$

Differentiate partially with respect to x,

$$\frac{\partial}{\partial x} \ln \frac{q}{p} = T'p - \frac{X''}{X'}.$$

Differentiate partially with respect to y,

$$\frac{\partial^2}{\partial x \partial y} \ln \frac{q}{p} = T''pq + T's$$

or

$$\frac{1}{pq} \frac{\partial^2}{\partial x \partial y} \ln \frac{q}{p} = T'' + T' \frac{s}{pq}.$$

Let

$$v = \frac{s}{pq} \quad \text{and} \quad u = \frac{1}{pq} \frac{\partial^2}{\partial x \partial y} \ln \frac{q}{p},$$

then

$$u = T'' + vT'. \tag{A.12}$$

Differentiate (A.12) partially with respect to x,

$$\frac{\partial u}{\partial x} = \frac{\partial (T'' + vT')p}{\partial z} + T' \frac{\partial v}{\partial x}. \tag{A.13}$$

Differentiate (A.12) partially with respect to y,

$$\frac{\partial u}{\partial y} = \frac{\partial (T'' + vT')q}{\partial z} + T' \frac{\partial v}{\partial y}. \tag{A.14}$$

From (A.13) and (A.14)

$$pq \frac{\partial (T'' + vT')}{\partial z} = p \left(\frac{\partial u}{\partial y} - T' \frac{\partial v}{\partial y} \right) = q \left(\frac{\partial u}{\partial x} - T' \frac{\partial v}{\partial x} \right)$$

giving

$$T' = \frac{q\frac{\partial u}{\partial x} - p\frac{\partial u}{\partial y}}{q\frac{\partial v}{\partial x} - p\frac{\partial v}{\partial y}}.$$

Setting $w = T'$ we have

$$\frac{\partial w}{\partial x} = T''p \quad \text{and} \quad \frac{\partial w}{\partial y} = T''q.$$

Substituting in (A.12)

$$\frac{1}{p}\frac{\partial w}{\partial x} = \frac{1}{q}\frac{\partial w}{\partial y} = u - vw$$

which are Lecornu's conditions.

APPENDIX B

Elements from Kellogg's Work on Anamorphosis

If

$$F(x,y,z) = P_1(x)R_1(y,z) + P_2(x)R_2(y,z) + P_3(x)R_3(y,z),$$

then $F(x,y,z)$ satisfies the ordinary homogeneous linear differential equation

$$\begin{vmatrix} F & P_1 & P_2 & P_3 \\ F_x & P_1' & P_2' & P_3' \\ F_{xx} & P_1'' & P_2'' & P_3'' \\ F_{xxx} & P_1''' & P_2''' & P_3''' \end{vmatrix} = 0.$$

The necessary and sufficient conditions for the existence of this equation are that the matrix

$$\begin{pmatrix} F & F_y & F_z & F_{yy} & F_{yz} & F_{zz} & F_{yyy} & F_{yyz} & F_{yzz} & F_{zzz} \\ F_x & F_{xy} & F_{xz} & F_{xyy} & F_{xyz} & F_{xzz} & F_{xyyy} & F_{xyyz} & F_{xyzz} & F_{xzzz} \\ F_{xx} & F_{xxy} & F_{xxz} & F_{xxyy} & F_{xxyz} & F_{xxzz} & F_{xxyyy} & F_{xxyyz} & F_{xxyzz} & F_{xxzzz} \\ F_{xxx} & F_{xxxy} & F_{xxxz} & FF_{xxxyy} & F_{xxxyz} & F_{xxxzz} & F_{xxxyyy} & F_{xxxyyz} & F_{xxxyzz} & F_{xxxzzz} \end{pmatrix}$$

be of rank less than 4 and the matrix

$$\begin{pmatrix} F & F_y & F_z & F_{yy} & F_{yz} & F_{zz} \\ F_x & F_{xy} & F_{xz} & F_{xyy} & F_{xyz} & F_{xzz} \\ F_{xx} & F_{xxy} & F_{xxz} & F_{xxyy} & F_{xxyz} & F_{xxzz} \end{pmatrix}$$

be of rank 3.

251

APPENDIX C

Best Functional Scale is Logarithmic

The following shows that the best functional scale is the logarithmic scale, assuming that the error is measured by relative error.

For a regular scale the scale factor L is constant and is given by

$$L = \frac{f(x_{n+1}) - f(x_n)}{x_{n+1} - x_n}.$$

If the scale is not regular the scale factor will vary with x. At some point x_1 the scale factor will be given by

$$L_{x_1} = \lim_{x \to x_1} \left(\frac{f(x) - f(x_1)}{x - x_1} \right) = f'(x_1).$$

The following two conditions are required of this scale:

i. $r = x^{-1} e_x$ is constant
ii. $m = L_x e_x$ is constant

where r is the relative error and e_x is the error in x. It follows that

$$\frac{m}{r} = \frac{L_x e_x}{e_x} x.$$

In other words,

$$x L_x = C$$

$$\text{so} \quad f'(x) = \frac{C}{x}$$

$$\text{and} \quad f(x) = C \ln x.$$

Therefore the best scale is a logarithmic scale.

Equivalence of Results of Steyskalova and Buck

Steyskalova's theorem states that, for an equation $F(x, y, z) = 0$ to be of the third nomographic order it is necessary and sufficient that the following six equations are such that each is a consequence of the other five:

$$F(x_0, y_1, z_1) = 0,$$
$$F(x_0, y_2, z_2) = 0,$$
$$F(x_1, y_2, z_3) = 0, \qquad\qquad \text{(D.1)}$$
$$F(x_2, y_1, z_3) = 0,$$
$$F(x_2, y_0, z_2) = 0,$$
$$\text{and} \quad F(x_1, y_0, z_1) = 0.$$

Since $F(x, y, z) = 0$ is of the third order it can be rewritten as $z = G(x, y)$. Taking particular pairs of the equations of (D.1), say

$$F(x_0, y_1, z_1) = 0 \quad \text{and} \quad F(x_1, y_0, z_1) = 0,$$

they can be expressed in the forms

$$z_1 = G(x_0, y_1) = G(x_1, y_0).$$

Steyskalova's result can now be stated in the following form:

If

$$G(x_0, y_1) = G(x_1, y_0) \quad \text{and} \quad G(x_0, y_2) = G(x_2, y_0),$$

then if

$$z_3 = G(x_1, y_2)$$

it follows that

$$z_3 = G(x_1, y_2) \quad \text{or} \quad G(x_2, y_1) = G(x_1, y_2).$$

Buck's result deals with the expression,

$$z = F(x, y) = f(\phi(x) + \psi(y)),$$

again an expression of the third nomographic order. His result states that for any a and c in the range of x and any b and d in the range of y, with $|a - c|$ and $|b - d|$ sufficiently small, there must exist x and y near a and b respectively such that,

$$F(a, b) = F(c, y),$$
$$F(c, d) = F(x, b),$$
$$\text{and} \quad F(a, d) = F(x, y).$$

The sufficiently small criterion apart, this is the essence of Steyskalova's result with, for example, a corresponding to x_0, c to x_1, x to x_2, y to y_0, b to y_1, and finally d to y_2.

Bibliography

[1] Harold John Allcock and J. Reginald Jones. *The Nomogram: The Theory and Practical Construction of Computation Charts.* Pitman, New York, 1932.

[2] Harold John Allcock, J. Reginald Jones, and J.G.L. Michel. *The Nomogram: The Theory and Practical Construction of Computation Charts.* Pitman, London, 1963.

[3] Georges-Jean-Baptiste-François Allix. *Explication d'un nouveau systéme de tarifs; ou, Nouvelle méthode pour trouver, in mesures métriques, sans aucun calcul, le poids des m'etaux en barres ou en feuilles ... la capacité des tonneaux, des vases cylindriques, etc.* Bachelier, Paris, 1840.

[4] Vladimir Igorevich Arnol'd. On functions of three variables. *Amer. Math. Soc. Translations*, 28(2):51–54, 1963.

[5] Jean Elie de Beaumont, Gabriel Leme, and Augustin-Louis Cauchy. Rapport sur un mémoire de M. Léon Lalanne, qui a pour objet la substitution de plans topographiques à des tables numériques à double entrée. *C. R. Académie Sci.*, pages 492–494, 1843.

[6] Paracelse Elie Désiré Bellencontre. *Table figurative du tir d'après Lombard.* Leneveu, Paris, 1830.

[7] Wilhelm Blaschke. Sui problemi fondamentali della Nomografia. *Rend. Mat. e Appl. (5)*, 15:46–52, 1956.

[8] Wilhelm Blaschke and Gerrit Bol. *Geometrie der Gewebe.* Springer-Verlag, Berlin, 1938.

[9] Farid Youssef Boulad. Application de la méthode des points alignés au tracé des paraboles de degré quelconque. *Annales des Ponts et Chaussées*, 22:255–268, 1906.

[10] Farid Youssef Boulad. Sur la disjonction des variables des équations nomographiquement rationnelles d'ordre supérieur. *C. R. Académie Sci.*, 150:379–382, 1910.

[11] Carl Benjamin Boyer. Early graphical solutions of polynomial equations. *Scripta Mathematica*, pages 5–19, 1945.

[12] Selig Brodetsky. *A First Course in nomography.* Pitman, New York, 1922.

[13] Philippe Buache. Essai de géographie physique. *Mem. Académie Roy. Sci.*, pages 399–416, 1752.

[14] R. Creighton Buck. Approximate complexity and functional representation. Technical report, Mathematics Research Centre, University of Wisconsin, July 1976.

[15] Donald S. L. Cardwell. *The Organization of Science in England.* Heinemann, London, 1977.

[16] V.A. Cherpasov and G.E. Džems-Levi. The calculation of approximate nomograms on a fast working machine. *Vychislitelnaya Matematika*, 4:160–161, 1959.

257

[17] Major W. H. Chippindall. Graphic solution for equations. *Professional Papers of the Corps of Royal Engineers*, 19:177–187, 1893.

[18] J. Clark. Théorie générale des abaques d'alignment de tout ordre. *Revue de Mécanique*, 21:321–335, October 1907.

[19] J. Clark. Théorie générale des abaques d'alignment de tout ordre. *Revue de Mécanique*, 21:575–585, December 1907.

[20] J. Clark. Théorie générale des abaques d'alignment de tout ordre. *Revue de Mécanique*, 22:236–253, March 1908.

[21] J. Clark. Théorie générale des abaques d'alignment de tout ordre. *Revue de Mécanique*, 22:451–472, May 1908.

[22] Carl Culmann. *Traité de Statique Graphique*, volume 1. Dunod, Paris, 1880.

[23] I.N. Denisyuk. Effective formulae of projective transformation and their application in the construction of empirical dependency. *Vychislitelnaya Matematika*, 4:162–166, 1959.

[24] I.N. Denisyuk. Some polynomials and the nomogram for their construction. *Vychislitelnaya Matematika*, 4:167–179, 1959.

[25] E. Deville. Abacus of the pole star. *Transactions of the Royal Society of Canada*, 12:3–11, 1906.

[26] Leonard Eugene Dickson. *Modern Algebraic Theories*. B.H. Sanborn, Chicago, 1930.

[27] Isidore Didion. Expériences sur la justesse comparée du tir des balles sphériques, plates at longues. *J. École Polytechnique*, 27:51–74, 1839.

[28] Surrey Directorate of Overseas Surveys, Tolworth. Nomograms for survey compututions. Technical report, Directorate of Overseas Surveys, Tolworth, Surrey, January 1973.

[29] Marcelin Du Carla. *Expression des nivellemens ou méthode nouvelle pour marquer rigoureusement sur les cartes terrestres et marines les hauteurs et les configurations du terrein*. L. Cellot, Paris, 1782.

[30] Sir Robert Dudley. *Dell'Arcano del Mare*. Giuseppe Cocchini, Florence, 1661.

[31] Ernest Duporcq. Sur la théorie des abaques à alignment. *C. R. Académie Sci.*, 127:265–268, 1898.

[32] G. E. Džems-Levi. A nomogram for the integral law of Student's distribution. *Teor. Veroyatnost. i Primenen.*, 1:272–274 (1 plate), 1956.

[33] G.E. Džems-Levi. Nomogram for projective transformation of scales and its application in the construction of nomograms. *Nomographic collection MGU*, page UNKNOWN, 1951.

[34] G.E. Džems-Levi. Normierte massausche determinanten und angenäherte konstruktion von nomogrammen. *Uch. Zap. Mosk. Gos. Univ.*, 163:133–136, 1952.

[35] G.E. Džems-Levi. The projective transformation of nomograms. *Uspekhi Mat. Nauk*, 7:147–151, 1952.

[36] G.E. Džems-Levi. On the problem of general anamorphosis. *Dokl. Akad. Nauk SSSR (N.S.)*, 113:258–260, 1957.

[37] G.E. Džems-Levi. On nomogramming of equations of 4th nomogrammic order. *Matematicheskii Sbornik*, 44(1):123–130, 1958.

[38] G.E. Džems-Levi. Einige allgemeine methoden der praktischen nomographie. *Vychislitelnaya Matematika*, 4:104–149, 1959.

[39] G.E. Džems-Levi. A nomographic method of proof of certain theorems. *Uch. Zap. Mosk. Gos. Univ.*, 186:249–252, 1959.

[40] G.E. Džems-Levi. Some central methods of practical nomography. *Vychislitelnaya Matematika*, 1959.

[41] G.E. Džems-Levi. Über funktionen, deren nomogramme eine vorgegebene resultatskala haben. *Vychislitelnaya Matematika*, 5:109–132, 1959.

[42] G.E. Džems-Levi. On the binary anamorphosis. Nomograf. Zbornik, Zbornik Prednasok III. celostat. Konf. Nomografii, Kosice 1970, 49-52 (1970)., 1970.

[43] G.E. Džems-Levi and D.L. Lipatova. Standardization of projective transformations. *Uch. Zap. Mosk. Gos. Univ.*, 161:235–240, 1956.

[44] Jean Baptiste Simon Fevre. *Traité du mouvement de translation des locomotives et recherches sur le frottement de roulement*. Fain et Thunot, Paris, 1844.

[45] G. Fontené. Formes réduites d'une relation triplement linéaire entre trois variables. *Nouvelles annales de mathématiques*, 19:494–498, 1900.

[46] Michael Friendly. Visions and Re-Visions of Charles Joseph Minard. *Journal of Educational and Behavioral Statistics*, 27(1):31–51, 2002.

[47] J.L. van Genderen. Nomograms for morphometric gravel analysis. *Sedimentary Geology*, 17:285–294, 1977.

[48] M. Gorodskii. Uchanie zapiski. *MGU*, 28, 1939.

[49] D. Gorrieri. *Résistance des poutres chargées*. Atti del Collegio degli Ingegneri ed Architetti, Bologne, 1895.

[50] Domenico Gorrieri. Résistance des poutres chargées. *Bull. del Collegio degli Ing. ed. Archit.*, 1896.

[51] Thomas Hakon Grönwall. Sur les équations entre trois variables représentables par des nomogrammes à points alignés. *Journal de Mathmatiques Pures et Appliquées*, 8:59–102, 1912.

[52] R.A. Hezlet, Capt R.K. Scale for the graphic calculation of deflection and angle of sight problems. *The Journal of the Royal Artillery*, 36(4):190–192, 1909.

[53] R.A. Hezlet, Capt R.K. The graphic representation of formulae. *The Journal of the Royal Artillery*, 36(10):457–470, 1910.

[54] R.A. Hezlet, Lt. Col. R.K. What is a nomogram? *The Journal of the Royal Artillery*, 45(10):330–333, 1920.

[55] R.K. Hezlet. *Nomography or the graphic representation of formulæ*. The Royal Artillery Institution, 1913.

[56] David Hilbert. Mathematische probleme. vortrag, gehalten auf dem internationalen mathematiker-congress zu paris 1900. *Gött. Nachr.*, pages 253–297, 1900.

[57] David Hilbert. Mathematical problems. *Bull. Amer. Math. Soc.*, 8:437–479, 1902.

[58] Ludwig Friedrich Kämtz, Charles Martins, Leon Lalanne, and Charles V. Walker. *A Complete Course of Meteorology*. H. Bailliére, 1845.

[59] O. D. Kellogg. Nomograms with points in alignment. *Z. Maths Phys*, 63:159–173, 1915.

[60] G.S. Khovanskii. Some problems of practical nomography. *Vychislitelnaya Matematika*, 4:3–103, 1959.

[61] Donald Ervin Knuth. *Computers & Typesetting, Volume A: The TeXbook*. Addison-Wesley, New York, 1986.

[62] Andrey Nikolaevich Kolmogorov. On the representation of continuous functions of several variables by superpositions of continuous functions of a smaller number of variables. *Amer. Math. Soc. Translations*, 17(2):369–373, 1961.

[63] Andrey Nikolaevich Kolmogorov. On the representation of continuous functions of many variables by superposition of continuous functions of one variable and addition. *Amer. Math. Soc. Translations*, 28(2):55–59, 1963.

[64] Léon-Louis Chretien Lalanne. Mémoire sur la substitution de plans topographiques à des tables numériques à double entrée. *C. R. Académie Sci.*, pages 1162–1164, 1843.

[65] Léon-Louis Chretien Lalanne. Mémoire sur les tables graphiques et sur la géométrie anamorphique appliquées à diverses questions qui se rattachent à l'art de l'ingénieur. *Annales des Ponts et Chaussées*, 11:1–69, 1846.

[66] Léon-Louis Chretien Lalanne. De l'emploi de la géométrie pour résoudre certaines questions de moyennes et de probabilités. *J. Mathématiques Pures et Appliquées*, 50:107–130, 1876.

[67] Léon-Louis Chretien Lalanne. *Graphic Methods for the Expression of Empirical or Mathematical Laws with Three Variables with their Application to Engineering and to the Solution of Numerical Equations of any Degree*, chapter 38, pages 346–415. National, Paris, 1880.

[68] Charles Lallemand. *Les abaques hexagonaux: Nouvelle méthode générale de calcul graphique, avec de nombreux exemples d'application*. Ministère des travaux publics, Comité du nivellement général de la France, Paris, 1885.

[69] Charles Lallemand. Sur une nouvelle méthod générale de calcul graphique au moyen des abaques hexagonaux. *C. R. Académie Sci.*, 102:816–819, 1886.

[70] Charles Lallemand. Nivellement de haute précision. In *Lever des Plans de Nivellement*. Librairie Polytechnique, Paris, 1889.

[71] Charles Lallemand. Sur la genèss et l'état actuel de le science des abaques. *C. R. Académie Sci.*, 102:82–88, 1922.

[72] Leslie Lamport. *LaTeX: A Document Preparation System*. Addison-Wesley, New York, 2nd edition, 1994.

[73] D.C. Lapteva. On the projective transformation of alignment nomograms with rectilinear solution scales. *Vychislitelnaya Matematika*, 4:150–158, 1959.

[74] Václav Láska. Zeitschrift für vermessungsween. 1905-1906.

[75] S. Lavington. *Early British Computers*. Manchester University Press, Manchester, 1980.

[76] L. Lecornu. Sur le problème de l'anamorphose. *C. R. Académie Sci.*, 102:813–816, 1922.

[77] Lieut du Genie Lelarge. Méthode telephotographiques. *Revue du génie militaire*, 32:193–230, September 1906.

[78] Lieut du Genie Lelarge. Méthode telephotographiques. *Revue du génie militaire*, 32:295–324, October 1906.

[79] George Margetts. *Margetts's horary tables : for shewing by inspection, the apparent time, from altitudes of the sun, moon, and stars, the latitude of a ship, & the azimuth, time, or altitude, corresponding with any celestial object*. Private, London, 1790.

[80] Wladimir Margoulis. Les abaques à transparent orienté. *C. R. Académie Sci.*, 174:1684–1686, 1922.

[81] Wladimir Margoulis. Sur la theorie générele de la représentation des equations au moyen d'éléments mobiles. *C. R. Académie Sci.*, 176:824–826, 1923.

[82] Junius Massau. Mémoire sur l'intégration graphique et ses applications. *Annales Assoc. d. Ingénieurs sortis d. Écoles Spéciales de Gand*, pages 70–102, 1884.

[83] August Ferdinand Möbius. La table graphique de multiplication. *J. Reine Angew. Math.*, 22:280, 1841.

[84] Molfino. *Rivista marittima*. Zanichelli, Bologna, 1896.

[85] Alexandre Magnus d' Obenheim. *Balistique: indication de quelques expériences propres à compléter la théorie du mouvement des projectiles de l'artillerie : précédée de l'analyse nécessaire e.* Levrault, Strasbourg, 1814.

[86] Alexandre Magnus d' Obenheim. *Mémoire contenant la théorie, la description et l'usage de la planchette du cannonnier: ouvrage dedie A.S.A.R. Monseigneur le Duc de Berry.* F.G. Levrault, Strasbourg, 1818.

[87] Maurice d' Ocagne. Étude de deux systémes simples de coordonnées tangentielles dans le plan. *Nouvelles Annales de Mathematiques*, 3:410–470, 1884.

[88] Maurice d' Ocagne. Procédé nouveau de calcul graphique. *Annales des Ponts et Chaussées*, 8:531–540, 1884.

[89] Maurice d' Ocagne. *Coordonnées parallèles et axiales. Méthode de transformation géométrique et procédé nouveau de calcul graphique déduits de la considération des coordonnées parallèles.* Gauthier-Villars, Paris, 1885.

[90] Maurice d' Ocagne. *Nomographie. Les calculs usuels effectués au moyen des abaques.* Gauthier-Villars, Paris, 1891.

[91] Maurice d' Ocagne. Nomography: On equations representable by three linear systems of isoplethé points. In *Mathematical papers read at the International Congress*, pages 258–271. American Mathematical Society, 1896.

[92] Maurice d' Ocagne. *Traité de Nomographie*. Gauthier-Villars, Paris, 1899.

[93] Maurice d' Ocagne. Sur la résolution nomographique de l'équation du septieme degré. *C. R. Académie Sci.*, pages 522–524, 1900.

[94] Maurice d' Ocagne. Sur les barycentres cycliques dans les courbes algébriques. *Bull. Soc. Math. France*, 30:83–91, 1902.

[95] Maurice d' Ocagne. Sur la résolution nomographique des triangles spheriques. *C. R. Académie Sci.*, 138:70–72, 1904.

[96] Maurice d' Ocagne. Sur la résolution nomographique générale des triangles sphériques. *Bulletin de la Soc. Math. de france*, 32:196–203, 1904.

[97] Maurice d' Ocagne. Sur un théorème de j. clark. *C. R. Académie Sci.*, 142:986–990, 1906.

[98] Maurice d' Ocagne. Les progres recents de la méthode nomographique des points alignes. *Rev. Gen. des Sci. Pure & Applique*, 16:392–395, 1907.

[99] Maurice d' Ocagne. Sur la représentation par points alignes de l'équation d'ordre nomographique 3 le plus generals. *C. R. Académie Sci.*, 144:190–192, 1907.

[100] Maurice d' Ocagne. Sur le répresentation de l'équation d'ordre nomographique 3 la plus générale par un nomogramme conique. *C. R. Académie Sci.*, 144:895–898, 1907.

[101] Maurice d' Ocagne. Sur le répresentation des équations d'ordre nomographique 4 à 3 et 4 variables. *C. R. Académie Sci.*, 144:1027–1030, 1907.

[102] Maurice d' Ocagne. *Cours de Géométrie Pure et Apliquée.* Gauthier-Villars, Paris, 1917.

[103] Maurice d' Ocagne. *Traité de Nomographie*. Gauthier-Villars, Paris, 2nd edition, 1921.

[104] Maurice d' Ocagne. Le calcul nomographie avant la nomographie. *Ann. Soc. Sci. de Bruxelles*, 2:55–66, 1926.

[105] Maurice d' Ocagne. Les archives nomographiques de l'école des ponts et chaussées. *Rev. gen. des sciences pures et apliquées*, 39:625–626, 1928.

[106] El general Diego Ollero y Caromna. *Nomografla balistica*. Écija, Segovia, 1903.

[107] Edward Otto. *Nomografia*. Państwowe Wydawnictwo Naukowe, Warsaw, 1956.

[108] Edward Otto. *Nomography*. Pergamon Press, New York, 1963.

[109] John B. Peddle. The construction of graphical charts. *The American Machinist*, May, September and November 1908.

[110] M.V. Pentkovsky. *Cours de Géométrie Pure et Appliquée*. Gostekhizdat, 1949.

[111] M.V. Pentkovsky. Skeletons of nomograms of equations of the third nomographic order. Technical report, AN USSR, 1953.

[112] M. Perret. Note sur quelques applications de la nomographie à l'astronomie nautique. *Annales Hydrographiques*, pages 170–199, 1904.

[113] G. Pesci. *Cenni di nomografia*. Tipografia di R. Giusti, Livorno, 2nd edition, 1901.

[114] Guillaume Piobert. Unknown. *Mémorial de l'artillerie*, 1826.

[115] Julius Plücker. Über eine neue art, in der analytischen geometrie puncte und curven durch gleichungen darzustellen. *J. Reine Angew. Math.*, 6:107–146, 1830.

[116] Louis Ézéchiel Pouchet. *Arithmétique linéaire*. Seyer, Rouen, 1795.

[117] René Poussin. *Diagrams for Egyptian Engineers*. Le Caire, Cairo, 1904.

[118] René Poussin. *Sur l'application des procédés graphiques aux calculs d'assurances*. L. Dulac, Paris, 1904.

[119] Guiliano Ricci. *La Nomografia*. Voghera, Rome, 1901.

[120] J. Rickard. A graphical procedure for computing the effective rate of interest. *Bulletin of the Institute of Mathematics and its Applications*, 6(10):228–229, 1980.

[121] Giuseppe Ronca. *Manuale del tiro*. Tipografia di R. Giusti, Livorno, 1901.

[122] Paul de Saint-Robert. De la resolution de certaines equations a trois variables. *Memorie della R. Academia di Torino*, 25:53–62, 1871.

[123] R.E. Sankey, Capt. C.E.P. Moving loads on military bridges. *Professional papers of the Corps of Royal Engineers*, 2(6):164–172, 1911.

[124] Scholfield. The use of logarithmic scales in plotting curves. *Minutes of Proceedings of the Institute of Civil Engineers*, 154:287–291, 1903.

[125] S.V. Smirnov. Uchenie zapiski. *Ivanovskogo Gos. ped. Institute*, 4, 1953.

[126] Rodolphe Soreau. Contribution à la théorie et aux applications de la nomographie. *Extrait des Mémoires de la Société des ingénieurs civls de France*, 54:191–243, 1901.

[127] Rodolphe Soreau. *Nomographie ou Traité des Abaques*. Chiron, Paris, 1921.

[128] Rodolphe Soreau. Pour servir a l'histoire de la nomographie. *Revue générale des sciences*, 33:518–523, 1922.

[129] T. Steyskalova. Elements of the theory of nets and its application in nomography. *Vychislitelnaya Matematika*, 4:173–183, 1959.

[130] Ellen Swanson, Arlene O'Sean, and Antoinette Schleyer. *Mathematics into Type*. American Mathematical Society, New York, 1999.

[131] Olry Terquem and Carl Friedrich Borkenstein. Extrait sur cartouches á balles du traité d'artillerie théorique et pratique. *Mémorial de l'artillerie*, 1:306–321, 1830.

[132] F. J. Vaes. Technische rekenplaten. *Ingenieur*, 19(7):322ff., May 1904.

[133] I.A. Vilner. *Nomograficheskii Sbornik*, 1951.

[134] Mieczyslaw Warmus. Nomographic functions. *Rozpravy Matematyczne*, 16:3–149, 1959.

[135] M. Wolff. *Diagrams for Egyptian Engineers*. Le Caire, Cairo, 1904.

Index

1601755R00148

Printed in Germany
by Amazon Distribution
GmbH, Leipzig